<u>365</u>
<u>Technical Writing Tips</u>

<u>Keith Johnson, M.S. Education, CSPO</u>

<u>Senior Technical Writer, Technology Author</u>

<u>Hollywood, Florida, USA</u>

Table of Contents

Prologue by Dave Lance, Senior Technical Writer

In a conference room with two engineers. Writing the specs. They can't decide on the right word, so they ask you. You come up with a good one. The perfect one.

You're the writer.

They get busy building whatever it is they're building. Could be software. Firmware. A network. A security system. A packet sniffer. A mobile speaker. You attend their meetings. You load their code onto your workstation. Roll a beta unit into your cubicle. Become the first user of this new thing. You read the specs for the help system they are building into it. Or you provide them a .chm file. Don't worry. You have time to learn all the acronyms. Read the marketing literature. Devour the websites. Study the specs and the quality assurance tests. Google and Wikipedia are good friends. You are assigned docs to write. Maybe DITA to develop. You apply all your JoAnn Hackos knowledge, perform a proper and considerate user analysis. You write your doc plan. You set out to build the docs.

You read this book. Again.

You are assigned to your engineers. Maybe you attend and participate in agile scrums every morning. You have milestones. You have the support of the team. Every day you attend too many meetings. Sometimes you do not attend enough meetings. You gather information. (Here is a tip. Always record meetings with subject matter experts. Get a good digital recorder. And let them know you are using it. Transcribe the file when you get back to your desk. Use headphones. Send the engineer or SME an email demonstrating you heard exactly what they said.) Ask them for a seat in their lab. Set up your own system. Test what you've found. Get it down plain. Use FrameMaker or some other tool to present it well.

Use Illustrator, (or one of Keith's freeware suggestions) to illustrate it properly. Teach yourself how to create isometric drawings. Give the reader an optimal (technical) experience. Get them on their way. And return to your craft. Maybe take up video. ActionScript is so 2010.

Note: As fifteen minutes will pass between the time I write these words and the time you read them, look hard at what's coming. Bitcoin. Blockchain. Artificial Intelligence. Applications that monitor user behavior. Context driven help content. The evolution of XML. The embrace of video content. Stay abreast of the "Wired" content. Follow thought leaders on LinkedIn and Twitter. Read a few blogs. Register for some online seminars.

This is the first technical writing book (that I know about) that has huge heart. Kindness. (Common Sense even, as maligned as that is these days.) It addresses the craft of technical writing. It supports you in the role of user advocate. As such, it is timeless. An instant classic. I am proud to invite you to study it. Enjoy!

Best Regards,

Dave Lance, Senior Technical Writer

http://www.davelance.com

Preface by David Lazarus, Senior Graphic Artist

As someone who has worked in the IT field for 14 years, I know well the importance of documentation. Among other titles, I have worked as a Help Desk Analyst, Desk-side Support Technician and Systems Administrator. *All of those roles rely on technical writing in some compacity.*

In fact, you could say that those who provide technical support are like mini Technical Writers because they document each ticket and often provide step-by-step instructions to users. In some cases, those instructions were written by Technical Writers and the person providing tech support is merely following a script. However, there are many technical issues that require thinking outside the box. In those cases, it is important for the person providing the technical support to also provide good documentation.

Keith's extensive experience as a Technical Writer as well as his positive attitude and sincerity have made this book both informative and a pleasure to read. I highly recommend this book to Business Analysts, Subject Matter Experts and aspiring Technical Writers alike. This book gives you both understanding of many aspects of technical writing and, perhaps, a new-found respect for Technical Writers like Keith.

Best Regards,

David Lazarus, Senior Graphic Artist

https://pixels.com/profiles/david-lazarus.html

Introduction by Balam Abello, Independent Author

Although many people think that, in the modern world, people are not reading much due to the vast availability of multimedia content such as YouTube, this is not true.

More than ever, millions of people the world over are reading and writing all the time. They write e-mails, they write Instant Messages (IMs), they write Twitter messages, Facebook posts, Blog content, and more.

On the other hand, there has been a steady decline in book reading throughout the last decade. This is partly due to the overwhelming number of new titles added each year plus competing media, as stated above.

However, at the career level, people are required to write all sorts of documents.

This is where "365 Technical Writing Tips" comes in handy. This book is an extensive compilation of "how-to" tips, which can bring light to "at-times" complex writing topics.

Best Regards,

Balam Abello, Independent Author

Greetings from Keith Johnson, Book Author

Greetings and welcome to *365 Technical Writing Tips!*

My name is Keith Johnson, and I have been working in the field of Technical Communications since 1992. Over the course of my career, I have worked almost exclusively as a Technical Writer, but have also served the IT marketplace periodically as a Software Trainer. This book can be used by anyone who must write a technical document for his/her organization – business, school, and otherwise. About half of the tips in this book are *writing-specific*. The other half are *insights, resources and strategies* that can be used to expedite your technical documents.

ISBNs

ISBN-13: 978-1721036196
ISBN-10: 1721036199

Disclaimer

The following book, *365 Technical Writing Tips*, is neither personal nor professional advice, in any way. This book is for reading enjoyment only.

By reading ahead, you agree to release the author from all personal and/or professional liabilities regarding changes in your life that may or may not be related to your interpretation and/or use of what has been presented in this book. Content in this book may be shared only with the written permission of its author.

Caveat Emptor. Caveat Lector. Thank You.

The Professional Endeavor of Technical Writing

"Technical Writing is the practice of compiling, understanding, and presenting complex, technical information that meets documentation standards specified by a business, organization, or industry-specific authoritative body." - Keith Johnson, Senior Technical Writer.

The following book features 365 Technical Writing Tips based on my twenty-plus year career supporting software developers, business analysts, and engineers. Some of the tips in this book are for novice writers (like the tips about Microsoft Office) and some are for advanced writers (like the tip about Microsoft Visual Studio). Still, I believe that most tips in this book can be utilized by just about everyone. OK, it is time to take the plunge! The next section is the main body of the book:

365 Technical Writing Tips

Enjoy!

Best Regards,

Keith

Keith Johnson, Author

July 1, 2018

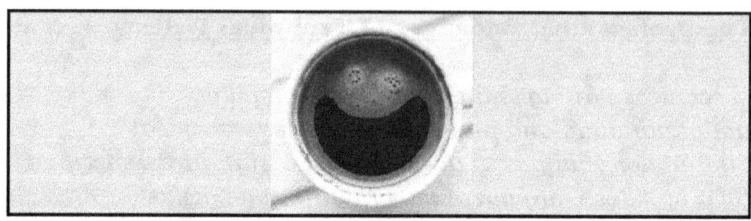

365 Technical Writing Tips

Tip 1: Acronyms (A) You Should Know

ACE = Access Control Entity
ADK = Application Development Kit
ADO = ActiveX Data Object
ADSI = Active Directory Service Interface
AIC = Application Integration Component
ANSI = American National Standards Institute
API = Application Programming Interface
APM = Advanced Power Management
ATM = Asynchronous Transfer Mode
AVI = Audio Video Interleaved

Tip 2: Acronyms (B) You Should Know

B2B = Business to Business
B2C = Business to Consumer
BBS = Bulletin Board System
BDC = Backup Domain Controller
BTFP = Broadcast File Transfer Protocol
BID = Board Interface Driver
BIOS = Basic Input Output System
BPS = Bits Per Second

Tip 3: Acronyms (C) You Should Know

CAD = Computer-Aided Design
CAM = Computer-Aided Manufacturing
CBT = Computer Based Training
CD = Compact Disc
CD-ROM = Compact Disc Read-Only Memory

CD-RW = Compact Disc Re-Writable
CGA = Color Graphics Adapter
CGI = Common Gateway Interface
CMC = Continuous Media Controller
CMS = Content Management System
COM = Component Object Model
CPU = Central Processing Unit
CRM = Customer Relations Management
CRT = Cathode Ray Tube
CSM = Certified Scrum Master
CSPO = Certified Scrum Product Owner
CSS = Cascading Style Sheet(s)
CTI = Computer Telephony Integration

Tip 4:   Acronyms (D) You Should Know

DAE = Data Access Engine
DAO = Data Access Object
DAT = Digital Audio Tape
DB = Data Base
DBCS = Double Byte Character Set
DBMS = Data Base Management System
DCOM = Distributed Component Object Model
DDE = Dynamic Data Exchange
DDI = Device Driver Interface
DDL = Data Definition Language
DDNS = Dynamic Domain Name System
DES = Data Encryption Standard
DFS = Distributed File System
DHTML = Dynamic Hyper Text Markup Language
DLL = Dynamic Link Library
DMOD = Dynamic Address Module
DNS = Domain Name System
DOM = Document Object Model
DOS = Disk Operating System
DRM = Digital Rights Management
DSN = Domain Server Name
DTD = Document Type Definition
DVD = Digital Versatile Disc

Tip 5:   Acronyms (E) You Should Know

ECC = Electronic Credit Card
EGA = Enhanced Graphics Adapter
EPUB = Electronic Publication
ERP = Enterprise Resource Planning

Tip 6:   Acronyms (F) You Should Know

FAT = File Allocation Table
FEK = File Encryption Key
FSD = Functional Standards Document
FTP = File Transfer Protocol

Tip 7:   Acronyms (G) You Should Know

GDI = Graphics Device Interface
GIF = Graphics Interchange Format
GPL = General Public License
GUI = Graphical User Interface

Tip 8:   Acronyms (H) You Should Know

HDD = Hard Disk Drive
HDMI = High-Definition Multimedia Interface
HKLM = HKey Local Machine
HTML = Hyper Text Markup Language
HTTP = Hyper Text Transfer Protocol
HTTPS = Hyper Text Transfer Protocol (Secure)

Tip 9:   Acronyms (I) You Should Know

IC = Integrated Circuit
ICP = Independent Content Provider
IDE = Integrated Device Electronics
IDE = Integrated Development Environment
IFS = Installable File System
IM = Instant Message, Instant Messaging
I/O= Input Output
IP = Internet Protocol
IPC = Inter Process Communication

ISP = Internet Service Provider
ISA = Industry Standard Architecture
ISDN = Integrated Services Digital Network
ISV = Independent Software Vendor

Tip 10: Acronyms (J) You Should Know

JPEG = Joint Photographic Experts Group
JSON = JavaScript Object Notation

Tip 11: Acronyms (K) You Should Know

KB = Kilo Byte
KBPS = Kilo Bytes Per Second
KBD = Keyboard
KPI = Key Performance Indicators

Tip 12: Acronyms (L) You Should Know

LAN = Local Area Network
LCD = Liquid Crystal Display
LED = Light Emitting Diode
LU = Logical Unit

Tip 13: Acronyms (M) You Should Know

MAC = Media Access Control
MCI = Media Control Interface
MDA = Monochrome Display Adapter
MDI = Multiple Document Interface
MFC = Microsoft Foundation Classes
MIDI = Musical Instrument Digital Interface
MIDL = Microsoft Interface Definition Language
MIME = Multipurpose Internet Mail Extension
MIPS = Millions of Instructions Per Second
MTA = Multi-Threaded Architecture
MTS = Microsoft Transaction Server

Tip 14: Acronyms (N) You Should Know

NAC = Network Adapter Card
NetBIOS = Network Basic Input Output System
NIC = Network Interface Card
NLB = Network Load Balancing

Tip 15: Acronyms (O) You Should Know

ODBC = Open Data Base Connectivity
OOM = Out of Memory
OOP = Object Oriented Programming
OS = Operating System

Tip 16: Acronyms (P) You Should Know

PC = Personal Computer
PDF = Portable Document Format
PFF = Printer File Format
PIF = Program Interface File
PIN = Personal Identification Number
PK = Primary Key
PROM = Programmable Read-Only Memory
PSU = Power Supply Unit

Tip 17: Acronyms (Q) You Should Know

QA = Quality Assurance
QC = Quality Control

Tip 18: Acronyms (R) You Should Know

RAID = Redundant Array of Independent Disks
RAM = Random Access Memory
RAS = Remote Access Server
RFI = Request for Information
RFP = Request for Proposal
RGB = Red-Green-Blue
RIFF = Resource Interchange File Format
RIP = Remote Imaging Protocol

RISC = Reduced Instruction Set Computer
ROM = Read-Only Memory
RPC = Remote Procedure Call

Tip 19: Acronyms (S) You Should Know

SaaS = Software as a Service
SBCS = Single Byte Character Set
SCSI = Small Computer System Interface
SDK = Software Development Kit
SDLC = Software Development Life Cycle
SGML = Standard Generalized Markup Language
SMS = Short Message Service
SMTP = Simple Mail Transfer Protocol
SNA = Systems Network Architecture
SQL = Structured Query Language

Tip 20: Acronyms (T) You Should Know

TIFF = Tagged Image File Format
TP = Transaction Processing
TDD = Test Driven Development

Tip 21: Acronyms (U) You Should Know

UI = User Interface
UML = Unified Modeling Language
UPS = Uninterruptible Power Supply
URL = Uniform Resource Locator
UX = User Experience

Tip 22: Acronyms (V) You Should Know

VM = Virtual Memory
VPN = Virtual Private Network
VR = Virtual Reality

Tip 23: Acronyms (W) You Should Know

WWW = World Wide Web
WYSIWYG = What You See Is What You Get
WAMP = Windows, Apache, MySQL, PHP

Tip 24: Acronyms (X) You Should Know

XHTML = Extensible Hypertext Markup Language
XML = Extensible Markup Language
XSL = Extensible Stylesheet Language

Tip 25: Acronyms (Y) You Should Know

Y2K = Year Two Thousand
YB = Yottabyte (One Septillion Bytes)

Tip 26: Acronyms (Z) You Should Know

ZAW = Zero Administration for Windows
ZB = Zettabyte (One Sextillion Bytes)

Tip 27: Admin vs. Non-Admin Documentation

As a general rule, I write for two distinct user audiences: *admins* (technical users) and *non-admins* (non-technical users). The tasks of each user group are easily defined. So, as you are writing user guides, for example, choose appropriate examples and presentation-strategies that will help each unique group accomplish such tasks.

Tip 28: Accuracy is a Writing Cornerstone

Productivity is a great thing, as long as your work is *accurate*. Sure, you can work faster and harder, but will you make a critical mistake that will then set you back even farther? *Take your time and be accurate.*

Tip 29: Active vs. Passive Voice

The active voice places the *doer* of an action at the beginning of the sentence. For example, let us consider the statement "the writer is composing a user guide." The *writer* is the doer of the action – composing – and this doer is placed before the verb, composing.

The passive voice is inferior because the *doer* is placed *after* the verb. One can invert this sentence into the passive voice by writing "a user guide is being composed by the *writer*." The *active voice is the best writing approach.* The *active voice* enables the reader to see, clearly, who/what is the *doer* of the action.

Tip 30: Adjectives Describe Nouns

In my path as a Technical Writer, *adjectives* serve as one of the *four and most essential* foundational pillars of English language sentences: *nouns, verbs, adverbs,* and *adjectives.*

Adjectives are words that describe nouns.

Examples: the beautiful hotel, the tall building, the long hallway, the historical home, the new computer, the old dictionary. The adjectives, here, are *beautiful, tall, long, historical, new,* and *old,* respectively.

Generally, adjectives are placed before the noun (when using active voice) so that by the time you see the noun, you are seeing it in light of the adjective. For technical documents, make sure that the adjectives you use are both *accurate* and *appropriate* for the nouns you are employing in your document.

Examples: the *popular* iPhone, the *Android* OS, the *extended* keyboard, the *multi-purpose* screwdriver, the *routing* switch, and the *electric* car.

Tip 31: Adverbs Describe Verbs and Adjectives

Adverbs describe both adjectives and verbs. Here are a few examples of *adverbs* in action:

The incredibly tall building is in downtown Miami.

The company unfairly decided to reduce sick days.

The tremendously generous technology billionaire donated one million dollars to a local charity.

Note: I will be addressing both nouns and verbs later in this book. Please consult with the book's Table of Contents to find these tips.

Tip 32: Adobe Acrobat

Technical Writers work with Adobe Acrobat for two main reasons: to *create* PDFs that can receive text-input through created fields and to *edit* PDFs to make them current and up-to-date. Most large corporations will grant their Technical Writers an *Adobe Acrobat* user license. If you are unable to obtain such a license, then I recommend its free equivalent: *Libre Office Draw*.

Tip 33: Adobe Captivate

Instructional Designers, and even some Technical Writers, use *Adobe Captivate* to produce interactive training presentations. If your company needs high-quality training and can afford *Captivate*, that is great. If your company (or you) cannot afford this program then seek out its free equivalent: *Adapt* (authoring tool).

Tip 34: Adobe Illustrator

Adobe Illustrator is a vector graphics computer program that allows users to compose and edit vector graphic images interactively and then save these images in one of the many popular vector graphic image file formats.

Illustrator is great for creating high-resolution images that can be used for marketing or even documentation purposes. If your company (or you) cannot afford Illustrator, then seek out its free equivalent: *Inkscape*.

Tip 35: Adobe InDesign

Adobe InDesign is popular software used for page layout that is superior to programs like Microsoft Word featuring only basic drawing/layout capabilities. If you cannot afford this software, then seek out its free and open-source equivalent: *Scribus*.

Tip 36: Adobe Photoshop

Adobe Photoshop is a *raster* (e.g. dot-matrix data structure) graphics editor developed and published by Adobe Systems for MacOS and Microsoft Windows. *Photoshop* is great for creating and improving images, graphs, charts, and more.

Photoshop is the *defacto* standard in the graphic artistry marketplace for both business and academia, but this software is expensive. So, if your organization (or you) cannot afford *Photoshop*, then seek out its free and open-source equivalent: *Gimp*.

Tip 37: Adobe Robohelp

Adobe Robohelp is my favorite HTML Help software system. You can create (1) compiled help (CHM) files that can reside on the user's desktop and (2) server-based HTML help systems that users can access through a web browser like Google Chrome. *Robohelp* is great because it allows users to find information quickly. If you cannot afford *Robohelp*, then seek out its free equivalent: *HelpNDoc*.

Tip 38: Affect vs. Effect

Affect is a verb meaning "to impact." *Effect* is a noun meaning "result."

Examples: If you leave the office temperature at 80 degrees Fahrenheit, this will *adversely affect* the server room. Please, keep the temperature at 60 degrees. The *side effect* of working seven days a week is stress which manifests as sickness and mistakes on deliverables.

Tip 39: Agile/Scrum

Agile/Scrum refers to software development strategies featuring *iterative* development. Here, initial system requirements and final solutions are attained through team collaboration. Agile methods feature a work ethic that encourages *team self-organization and accountability*. *Scrum* development processes are based on the *Agile Manifesto*, created by Jeff Sutherland. *Scrum* is a group approach to software development where work is performed in two or three-week sprints, and there are different types of meetings conducted to reinforce team productivity, task accountability, and code output.

Tip 40: Agree/Disagree with your Manager?

In the professional world, you will eventually reach a "crossroads" moment where a supervisor or manager will have something negative to say about your work. If my technical document is *incorrect*, I accept this feedback and immediately *correct* the document. If a supervisor or managers approves my document, I say "thanks" and ask why he or she has nothing to say. If you ever *disagree* with your supervisor, find out specifics (e.g. details) and then *address each individual detail separately*. Then, make your necessary edits and corrections.

Tip 41: Altruism and Empowerment

When we treat others with equal respect and dignity, then we are practicing *altruism*. As a Technical Writer, I frequently need to obtain highly technical and complex information from others like Subject Matter Experts (SMEs). Through *altruism* and caring for people above and beyond their immediate professional roles, I usually get the positive help/information/reciprocation that I need. Treating others equally and with *dignity and kindness* is something that all Technical Communicators must learn. *Altruism* empowers both the giver and receiver.

Tip 42: Among vs. Between

Among is a preposition implying "in the company of" or "in the presence of." *Between*, however, is a preposition implying "located between two determined points." When the points or entities are specified, use *between*.

Examples: *Between* the two workstations, there is a printer. When such points or entities are not specified, then use *among*. *Among* his ten paper choices, he chose the topic of *Vedic Mathematics*. Techies might say "*among* these Internet providers, we favor Provider B." Techies might also say "the server room is located *between* the AC room and the HR suite."

Tip 43: Antivirus Software

In today's sophisticated high-tech world, you need to do as much as you can to be safe, especially those who work with PCs and the Internet on a daily basis. The following two antivirus programs have kept me safe on local intranets and the web for many years: *Advanced System Care* (iObit) and *Windows Defender* (Microsoft).

Tip 44: API Documentation

API stands for *Application Programming Interface. API* is used to describe how an outside system communicates and interacts with an internal system.

Generally, when an outside system communicates with your company's system, there are *asynchronous* communications (one-way only, no response) and/or *synchronous* communications (two-way, response of some kind given). *API documentation* should cover the details of all such communications, both asynchronous and synchronous, on both business (high) and detail (code) levels. Now, system A and system B can connect.

Tip 45: Appear vs. Display

The word *most-shunned* by one of my Technical Writing mentors is *appear*. He once told me that *appear* is a "hocus-pocus" word used for magic tricks.

The magician will now make a rabbit <u>appear</u> from an empty hat. This language is not appropriate for technical documents.

You can say the word *display* instead of *appear*. For example, "the web page refreshes and *displays* your account information." Also, "the IDE compiles the code and then *displays* the output app."

Tip 46: Applying Business Rules

Business Rules are an aspect of the software development process that indirectly empower Technical Writers. For example, in your typical nTier (e.g. multi-tier) system (Database, Business Rules, Code, User Interface), it is *Business Rules* that allow the Technical Writer to have a definitive sense of boundary regarding where functionality starts/ends. *Business Rules* define the *scope* of a system.

Tip 47: Archive vs. Database

A database is more than just one aggregation of information. A database can also be an entire platform or group of platforms that host information for a specific organization or enterprise. *Archived data* is located in one specific database section. You have to go to a specific part of the database to access this *archived information.*

Tip 48: Arrive Alive

Florida State Troopers have a sticker on both the front and rear of their cruisers that says *arrive alive.* In other words, *slow down* and *drive carefully.* Even a slight physical injury sustained from an auto accident will negatively affect your work. Please, drive safely.

Tip 49: Artificial Intelligence (AI)

Technical Writers should embrace Artificial Intelligence (AI). Great AI systems are here to support all kinds of intensive research, in many fields, including Business Operations and Systems Analysis. Through AI, Technical Writers will be able to find answers to complex questions and information critical to technical documents they are creating. Here are three great AI projects which I have been following recently (IBM, Google, Microsoft):

IBM Watson
https://www.ibm.com/watson

Google AI
https://ai.google

Microsoft AI
https://www.microsoft.com/en-us/ai/default.aspx

Tip 50: Ask the Right Questions

Technical Writers need to prepare solid questions before meeting with Subject Matter Experts (SMEs), who are experienced professionals able to shed light on a technical topic. A solid list of "right questions" will propel the meeting forward in a way that will produce information necessary for new technical documents. How do you ask the right questions? Seek out the document *stakeholder* (the person for whom you are writing the document) and ask him/her to send you a list of questions that the business/organization wants answered.

Tip 51: Aspire vs. Inspire

Aspire and *Inspire* are two verbs that are similar but, in the end, different. *Aspire* means "to dream, to seek" while *inspire* means "to infuse, to motivate." *Aspire* is an intransitive verb, so it has no object. "I aspire to obtain a PhD in Education." *Inspire* is a transitive verb with an object. "The teachings of the late Dr. Wayne Dyer continue to inspire me greatly."

Tip 52: Audacity Audio Software

If you need a *free audio/recording software tool* that can record and playback sound in full stereo capacity, look no further than *Audacity*. I have been using this software for many years. You can record short meetings, short conversations, and more. It is a great meeting assistant as you gather information and conduct interviews where information flows rapidly.

Here is its web address: www.audacityteam.org.

Tip 53: Audience

If you went onto the LinkedIn business networking website and asked its community members "what is your top consideration when writing a document?", I am sure that those responding would most likely say "the audience". The entire purpose of writing a technical document is to *serve a specific audience*. Remember...

That [reading] *audience* might be executives.

That [reading] *audience* might be internal staff.

That [reading] *audience* might be app subscribers.

Audience is my main "mantra" as a Tech Writer.

Tip 54: Authoritative Writing

When you are writing a technical document, you need to establish yourself as a *thought leader*. This is a key cornerstone of what I call *authoritative writing*. In a given system, there are many ways to accomplish key tasks. Technical Writers need to produce high-quality text; then readers will follow your lead.

Tip 55: B2B vs. B2C

B2B means "business to business." Perhaps you are writing an API guide for a programmer who is creating an app that will interact with your company's system. So, your API guide, here, is a *B2B* document.

B2C, however, stands for "business to consumer." Perhaps the guide you are writing is for an anti-virus app that your company has created. The typical user is a regular individual. In this case, your guide is a *B2C* document.

Tip 56: Back Up Files Daily

Most Technical Writers work in some type of corporate cloud environment so "file backup" is handled by the company's IT Department. However, if you work locally on a laptop or desktop PC, *back up your work daily.*

Tip 57: Be Proud of Yourself, Always

For the most part, people become proud of themselves *after* achieving a goal. So, this means that the daily journey become arduous and unpleasant. This is not the way to go. Technical documents can take weeks to write. Each time you complete a major section, go out and take a break. Smile. *Stay proud/confident at each step.*

Tip 58: Binary: Zeros and Ones (Factoid)

Computers operate in binary and this means that they store data and information and perform calculations and operations *using only zeros and ones.* So, binary is a system where the entire alphanumerical system that we take for granted has a fundamental "base 2" representation in binary code. Here are some examples:

Decimal 0 = Binary 0; Decimal 1 = Binary 1; Decimal 2 = Binary 10; Decimal 3 = Binary 11; Decimal 4 = Binary 100; Decimal 5 = Binary 101; Decimal 6 = Binary 110; Decimal 7 = Binary 111; Decimal 8 = Binary 1000; Decimal 9 = Binary 1001; and Decimal 10 = Binary 1010.

Tip 59: Block Paragraphs Recommended

In block paragraphs, you do not indent on the first line of new paragraphs. Rather, you create a visible space between paragraphs. This presentation approach is helpful to readers as they slowly process new information, *paragraph by paragraph.*

Tip 60: Blogging is Cool

For those of you who really enjoy writing, I endorse blogging. Blogging allows you to establish yourself as a Subject Matter Expert (SME) where you can obtain sure and steady recognition and even income through peer recognition, advertising, and content patronage.

Tip 61: Bookboon

Bookboon (www.bookboon.com) is a "books" website from the UK that features hundreds (if not thousands) of high-quality business, technical, and motivational books. Check out this site; it features great e-books.

Tip 62: Born to Serve

Not everyone will agree with me on this point, but I sincerely believe that we are alive in this life in order to *serve* our fellow human beings. For me, Technical Writing has been a powerful way for me to learn complex technical systems and then write user guides that *empower others* as users of said systems. Serving others is a path to true happiness and contentment.

Tip 63: Business at the Speed of Thought

In 1999, Bill Gates and Collins Hemingway wrote a book called "Business at the Speed of Thought." *Please read this amazing book.* This book is about today!

Tip 64: Business Jargon

Business Jargon means "special business words and terms" that are used by certain groups that enable them to communicate more effectively. Tech Writers need to learn business/tech jargon for their tech docs.

Tip 65: Chat and Productivity

There are some departments, like customer service or technical support, where "chat" is an integral part of the working day. There are other professions, like computer programming and technical writing, where interruptions can become obstacles to productivity and completing complex deliverables. *Chat apps* empower you such that you remain active and informed at the same time.

Tip 66: Camtasia Studio

If you have to produce a video to train employees or workers, you can use *Camtasia Studio* by TechSmith. The software is not free, but it is reasonably priced and it is very easy to master. *Camtasia Studio* rocks!

Tip 67: Caution about Contradictions

Technical Writing is a professional writing style where one documents technical information and processes. For this reason, it is imperative that you *not contradict* yourself. Proofread your work carefully to make sure that your facts and conclusions are *accurate and consistent.*

Tip 68: Choosing the Document's Direction

I encourage you to work with both your supervisor and SMEs. Get management and SMEs to agree, side-by-side. Confirm this new agreement via email. Now you can begin writing with confidence.

Tip 69: Choose Your Battles Wisely

Work *with your supervisor* to determine what is most important. If people come to your desk, send them to your supervisor's desk and their request will be inserted into your work pipeline, appropriately.

Tip 70: Circle vs. Encircle

To "circle" something is to travel around it ... on a certain path, while to "encircle" means to surround something.

Example: I will *circle* the word on the paper; the army battalion will *encircle* the enemy forcing its surrender.

Tip 71: Click vs. Select

Click is a command you cite regarding a specific object that users must "call into action" in a software application. The user might have to *click* on a button or a radio button or a check box. *Select,* on the other hand, is a command you give to readers when there is a *choice* to be made. For example, in a drop-down menu, the user must *select* between three visible options.

Tip 72: Code by Charles Petzold

If you work in the field of IT or aspire to do so, then there is a book you absolutely must read, in my opinion. That book is *Code,* written by Charles Petzold, Microsoft's greatest Windows developer of all times. Great read!

Tip 73: Common Sense by Thomas Paine

I also encourage you to read *Common Sense* by Thomas Paine. The words of this United States Founding Father will inspire you. Technical Writers must apply *common sense,* daily, toward their writing assignments.

Tip 74: Computer Programming

Technical Writers do not need to know how to write code. Still, if you can learn some computer code, I recommend it. Top languages to learn today include Java, Python, and Microsoft C#.

Tip 75: Configuration vs. Customization

From the vantage point of software deployment, *configuration* refers to setting up a software program using what is provided *out of the box*. If new code needs to be created to enhance the existing system, however, then this is the *customization* process. Technical Writers should understand this distinction.

Tip 76: Conjunctions, Linking Ideas

In the English language, the most common conjunctions are: *for, and, nor, but, or, yet,* and *so.*

Examples: This document has been written <u>for</u> the CEO <u>and</u> for the CTO. That programming methodology is neither efficient <u>nor</u> practical. I will check with the IT manager, <u>but</u> he is usually not available. I proofread the document; <u>yet</u> the analyst found a few errors. The code failed; <u>so,</u> I will edit it before retesting it.

Tip 77: Consistency in Content Presentation

Consistency is what allows readers to mentally expedite the understanding your document. Not everyone is excited about reading a technical document. So, by being *consistent* in your presentation, you indirectly encourage readers to continue along until they finish.

Tip 78: Convey vs. Manage Information

In my professional opinion, Technical Writers *not only* convey (express) information *but also* manage (edit) information. Rarely is there a document that does not go through the revision process. Documentation changes in the same way that code and business processes change. Documentation needs to keep pace with these two major aspects of business.

Tip 79: Corroboration vs. Collaboration

Technical Writers will find themselves using these two words from time to time. First, *corroborate* is to confirm and give supporting information regarding a matter. Second, *collaborate* is to work together and work jointly. Examples: The evidence *corroborates* what the witness said about the event. The team members *collaborated* to get the technical document done before Friday.

Tip 80: Courage Required for Tech Docs

One of the hardest parts of being a Technical Writer is being asked to write complex documents. *That is why you are there!* Your manager cannot write it. SMEs cannot write it. So, you are tasked with the document. What is going to help you the most is the virtue of *courage.*

Tip 81: Continuous Process Improvement

Technical Writers produce documents that help staff to become more productive. Tech Writers help others to learn and grow with their organizations. Tech Writers are an active part of *Continuous Process Improvement (CPI).*

Tip 82: Craft

The Cambridge English Dictionary defines "craft" as a noun meaning "a skill in how to do or make something, or a job or activity needing such skill." Is Technical Writing a craft? In my professional opinion, *yes*, it is. The practice of writing a user guide is a technical *skill*, without a doubt.

Tip 83: CreateSpace

For those of you who might venture, one day, into publications as an author, I highly recommend Amazon's self-publishing company, "CreateSpace."

Tip 84: Creating Document Templates

I recommend the use of document templates regardless of your working situation. Whether you work by yourself or work in a large organization, document templates will help you to *expedite* final deliverables and *show* others you are capable of creating and putting out documents with a truly consistent look and feel.

Tip 85: Creating Section Breaks

A page break allows only for a physical separation of text and/or content from one page to another. When you call into service a *Section Break*, you are granting yourself the opportunity to also change page size, page orientation, new margins, and more. Recently, I had to create a technical document and needed to insert a table that had *many columns*. So, I used the "Section Break" feature in Microsoft Word. I was able to present the table with pages using the *Landscape* view (as opposed to *Portrait* view, which was used for the rest of the doc). Success!

Tip 86: Creating Tool Tips

Some of you in the Technical Writing field will end up working for large corporations with a sophisticated presentation of important technical documentation. The organization will most likely be working with HTML Help in some way and you might be asked to create something called "tool tips". This is where the user hovers his/her mouse/cursor over an object on the webpage and a tiny pop-up window displays with text/info about the object. Most likely, such a large corporation will already have in place a systematic approach to creating tool tips. But, if you are the first to take on this task, there is a reliable and free software you can use, called *OpenTip*. Its web address is www.opentip.org. Good luck!

Tip 87: Crediting Information Sources

All information comes from a source. It could be a Subject Matter Expert (SME). It could be an old IBM book. It could be a former professor. It could be a famous scientist. *You, as a Technical Writer, are obliged to reveal your sources in your document.* You need to create either a References or Bibliography section to make sure you are following Composition Best Practices. My favorite book on correct source citation is the *Chicago Manual of Style*.

Tip 88: Customer Relations

Technical Writers often support Sales and Customer Service teams by writing detailed and insightful technical documentation (especially *how-to* guides). Get to know your Customer Relations co-workers. They will share valuable information that you can use in your docs.

Tip 89: Data (Factoid)

There are three major classes of data types: *primitive* (machine, Boolean, numeric), *composite* (enumerations, strings/text types), and *others* (pointers and references, & function types). Technical Writers, please note.

Tip 90: Data Today

Data, in the world of Information Technology, is a term used to describe units of information that are stored as records in an electronic records-warehouse called a database. Today, we have relational databases like *My SQL, Microsoft SQL Server,* and *Oracle,* which are semi-intelligent software systems that store data and are able to respond to complex user requests for data, especially in one's quest to "generate" meaningful reports.

Tip 91: DITA (Factoid)

The acronym DITA stands for "Darwin Information Typing Architecture". DITA is an XML (Extensible Markup Language) data model used for both authoring and publishing information. DITA was created, originally, by IBM because it wanted to be able to reuse documentation content as software and hardware were improved, expanded upon, and released. DITA is a huge topic, so let me point you to a webpage that can tell you more:

http://docs.oasis-open.org/dita/v1.2/cd03/spec/DITA1.2-spec.pdf

Tip 92: Documenting a Database

Work with your Database Administrator (DBA) to determine the best tool or best set of tools to document your database. This is a huge task ... thanks.

Tip 93: DOS (Factoid)

Before Microsoft Windows came into being, PC users had to work with a *line-command* Disk Operating System (DOS), like MS-DOS. If you go to the following Wikipedia webpage, you will see the many different commands used to open files, search for files, view a batch file, and more.

https://en.wikipedia.org/wiki/List_of_DOS_commands

Tech Writers should know basic MS-DOS commands.

Tip 94: Doc Collaboration (Factoid)

There are times where a document will require more than just one author. There are tools now available in *Microsoft Word, Google Docs,* and other leading word processing platforms that enable for a document owner to authorize others to *collaborate* as co-document authors.

Tip 95: Frustrated Users

Our business and educational worlds, today, are sometimes highly dysfunctional. Mandates are often given to workers and students without a proper infrastructure to support their tech activities, especially when it comes to working with new computer software. The final product does not work as it should. *So, the result is frustrated end users.* The best businesses, organizations, and enterprises do not dismiss but rather *acknowledge frustrated users.* They collect such feedback and data and present this info to Development and/or Business teams for remediation.

Tip 96: Document Design

Technical Writers do more than just "compose" technical text. They also "design" documents. Technical Writers must determine the different sections and sub-sections of their technical content presentation. I have had people from different departments seek my assistance from time to time when it comes time to document design. Make sure such design is done in accordance with management input, oversight, and approval.

Tip 97: Document Direction and Flow

Documentation, in my professional opinion, can be written in two fundamental ways: *summary* and *process.*

If I have to document a pipeline process for a bank or financial lender from the initial customer call through underwriting and funding, for example, I am writing a *process* document. If I have to itemize all the parts of a helicopter, for example, with major assemblies and sub-assemblies, then I am writing a *summary* document.

Tip 98: Document Distribution

Over the past several months, through Office 365, I have seen great improvements to Microsoft SharePoint. You no longer need to distribute documents the old-fashioned way (through email attachments). Now, you can create a SharePoint site, for example, grant user access, and when you post a document to that SharePoint site, designated users gain immediate document access.

Tip 99: Document Duplication

One of the biggest challenges to Information Technology departments, these days, is security. Hackers sometimes breach systems not just to see what is there but also to try to destroy data. Regardless of whether you are cloud-based or local-machine/device based, duplicating your documents ensures the *continuance* of your work.

Tip 100: Document Review Process

Over the course of my Technical Writing career, there are *two individuals* whose opinion I procure adamantly: SME (content) and management (style). *Between SMEs and my supervisor, my document receives its needed review. It can now become a quality deliverable for the organization.*

Tip 101: Document Scope

One of the best ways to ensure that your document does not become too big is to utilize internal document links and references to other documents. This way, you can stay *focused* on your most important topic/topics.

Tip 102: Document Storage

There are many approaches to document storage. Still, I work most efficiently in the following way: *First,* I like to work locally. Perhaps there is an Internet connection problem or even a power problem. This is not a problem because I am working on a laptop with a local word processing application; and my laptop has battery power for six hours. *Second*, I like to back up my work, on a regular basis, to my Cloud account. *Third*, I back up my work to a USB-type external hard drive that I store in a secure and cool place. In this way, I have *three* copies of my work in totally unique locations (local, Cloud account, secure external hard drive).

Tip 103: Documentation and Migration

From 2008 - 2014, I worked for a well-known electronic payments software development company in Miami Beach, Florida. One of my tasks during this time was to document a system that was about to be retired as the company *migrated* to a new system. So, I had to document, for historical and legal purposes, the "retiring" system. It took me about six months to complete this massive document. Many times, as new systems come into play, there is great concern about writing user guides for the new system for employees/contractors to get up to speed/productivity. Still, remember that migration is a slow process and many times you have to look back for specific reasons. If you have solid documentation on the "retiring" system, you are in good standing. You will be able to successfully compare how something was done in the old system against how it is done in the new system.

Tip 104: Documentation and Training

Over the course of my Technical Writing career, I have discovered firsthand that Software Trainers and Technical Writers make for great professional colleagues. Sometimes a Tech Writer discovers a software functionality that the Trainer needs to learn. Sometimes the Trainer is briefed on a new functionality that the Tech Writer did not see in the system's formal requirements documentation. Both are on parallel paths. Consistent communication keeps both Tech Writers and Trainers "in the know."

Tip 105: Documentation as Core Asset

Change is the essence of life; and, at companies, people (employees and contractors) come and go. A company that invests in its system documentation will endure such changes. If a major player like a DBA leaves, you do not want the organization to go into panic mode. You want to be able to refer workers to already-written and securely-stored documentation that will help the organization to move ahead with confidence. Documentation matters.

Tip 106: Documentation as Permanence

Documentation can be called a permanent solution to the issue of change in the business world. *Here is why.* A business can cite specific documents as a way of defining its business activity, scope, revenue model, infrastructure, and more. This is a very important legality that should not be ignored. Technical Writers should be aware of this important truth and work accordingly to ensure their company's well-being in this respect.

Tip 107. Documentation as Orientation

I recently had to write an *Onboarding* guide for new employees and contractors. I originally composed this document because I was new to the organization and so I needed this information to support my role. So, I had to discover the jargon of the business, its revenue model, its personnel structure, and its business direction. I even created a *Glossary of Terms* for this guide. Many times, employees and contractors perform their daily tasks and have absolutely no idea about what the business is doing outside of their department. Employees and contractors who are *well-oriented* through informative documentation can become fully productive. Write to orient!

Tip 108: Documentation for Aerospace

AS9100 is the "defacto" US Aerospace Industry standard based on International Standards Organization (ISO) 9001 requirements that have been adapted to Aerospace. Businesses in this industry need to have rock-solid documentation that passes annual AS9100 reviews. Auditors are allowed to review R&D projects as well as products currently sold in the marketplace. If you work in this industry as a Technical Writer, be prepared for some long work hours.

Tip 109: Documentation for Audits

If you are in an industry where there is government regulation, then official corporate documentation is the first item auditors want to see. Such documentation answers auditor questions/concerns about en-route business, revenue models, and more. Take notice of this fact, Technical Writers.

Tip 110: Documentation for Finance

FINRA (Financial Industry Regulatory Authority) is the "defacto" organization that ensures the integrity of US financial markets. Technical Writers in this space might find themselves working with Report Writers, Business Intelligence (BI) professionals, DBAs (Database Administrators), CPAs, and CMAs. Technical Writers need to understand basic economics and accountancy to excel in the Financial Services industry.

Tip 111: Documentation for Sales

I have always enjoyed working with Sales and Marketing personnel. They are on the front lines and appreciate documentation support. Enjoy such writing tasks!

Tip 112: Documentation Transparency

The greatest asset regarding documentation transparency is an official and enforced "style" guide. In this guide, all professionals who write documents for an organization will be required to use *specific templates and presentation mandates* so that all documents seem as if they have been written by the same person (but they have been written by different people from different departments).

Tip 113: Documenting Source Code

Today, most IDEs (Integrated Development Environments) have tools that software developers can use to document their written code. An example that comes to mind is *Sandcastle* for Microsoft's Visual Studio IDE. There is also a great open-source code documentation tool called "Doxygen" that allows you to document code written in major languages like Java and C#. Technical Writers who learn languages like Java/C# can enhance and build upon IDE-generated code documentation.

Tip 114. Documenting a Web Portal

There are two important documents for a web portal: its requirements document and its user guide. The requirements document is created to help developers and quality assurance staff to stay on track regarding functionality. The user guide is normally written by a Tech Writer and shows users how to navigate the portal.

Tip 115: Documenting your Progress

A technical document can be a long journey. So, documenting your progress reminds you that you are making progress, even though it feels like you are still driving through the desert. Keep going!

Tip 116: Does Tech Writing Equal Tech Comm?

Technical Writing is a subdomain of Technical Communications. Other subsets/subdomains include technical training, technical instruction, and instructional design, just to name a few.

Tip 117: Editing, Proofreading

Technical Writers need to be able to edit copy. Whether you are writing articles, user guides, internal docs, external docs, marketing pamphlets, or training manuals, you need to be able to *revisit* your own work from the *editorial* standpoint. This practice will ensure that you have cohesive sentences, sensible paragraphs, clear formatting, and accuracy in your final document.

Tip 118: Education is Life

John Dewey (famous American philosopher) once said that "Education is life". Technical Writers are *educators* through their written documentation. Truth!

Tip 119: Efficacy vs. Efficiency

Efficacy is "the power or capacity to produce a desired result." Efficiency, on the other hand, is "the ability to act in a way that produces a minimal amount of waste."

Examples: The *efficacy* of the new jet engine during liftoff was stunning; within seconds we not only lifted off but were at full altitude. The *efficiency* of the student's invention proved to be formidable: it consumed one pound of trash and yielded only one teaspoon of liquid waste. Its patent was soon purchased by a famous billionaire.

Tip 120: Enterprise

The term *enterprise* refers to either a business or an activity of a business. Technical Writers should care deeply about the *enterprise* they serve. Technical Writers know that their documents enable and empower readers to use the *enterprise's* products/services well and this promotes the *enterprise's* reputation and value.

Tip 121: Error

I sincerely believe that error is natural part of the human experience. Error, even though it is disruptive, is an essential part of the creative and evolutionary process. By allowing technology to make decisions for us, our ability to think creatively will suffer. Many managers today are intolerant regarding worker error. Documentation and system source code can both have errors when there is miscommunication between BAs, SMEs, developers, and the project's Tech Writers. What is the solution?

Solid team communications allow for documentation (and source code) error minimization and elimination.

Tip 122: Email Communications

Even though text messaging is prevalent in today's modern world, I still believe in working with traditional *email communications* when it comes to business matters. You have a way to verify previous communications, emails sent, emails received, and more. Email is still a trustworthy business communications tool. Technical Writers should always confirm important information to co-workers and management through email. In this way, everyone is on the same page, always.

Tip 123: Employee vs. Contractor

Over the course of my career, I have worked as both an employee and contractor. Both engagements can be positive work experiences. When you are an employee of an organization, you feel close to all workers and the company itself, as a business entity.

As a contractor, you typically get paid per hour and do not have to work "free" for clients beyond regular work hours. Tech Writers need to choose work status selectively.

Tip 124: End User

From the Microsoft Writing Style Guide:

"Don't write or say 'end user'. Avoid 'user' when you can. Use, rather, *audience, customer, person, people, employee, coworker,* or *you* instead. It is OK to use 'user' in context for developers to distinguish the technology developer from the technology user. It is also OK to use 'user' in context for technology professionals to distinguish the system administrator from the system 'users'."

All Technical Writers should bookmark this tip!

Tip 125: Exercise on a Regular Basis

Technical Writers often sit at their workstations for long periods of time writing, emailing, speaking/engaging on the phone, and researching information. It is important for all those in Technical Communications to exercise on a regular basis to maintain proper circulation and heart health.

Please consult with your doctor to discover the best way for you to exercise on a regular basis. Your health matters!

Tip 126: Experience

All companies want experienced Technical Writers. Many connections at LinkedIn ask me how they can acquire experience in this field. I tell them to (1) work either as an intern at an established company or (2) volunteer toward an open source software development project.

With a basic Technical Writing skillset established through real-world work experiences, you can then approach the technology marketplace confidently.

Tip 127: File Naming

At home, I use the following approach to naming files: content_YYYYMMDD.extension. Naming files with this approach gives me the chance to search not only by name but also by date and file extension. At work, however, I follow file-naming protocols that conform with established business protocols. Ask your manager the best way to name/manage your files using company standards.

Tip 128: Flex Hours

If there is a way that you can obtain flexible working hours, make your move. Being at your desk at your workstation for a solid nine to ten hours a day (e.g. 8am to 6pm) cranking out documentation is tough. It can be tiring. Flex hours give you the chance to either work from home periodically or break up your work day so that you can exercise and/or take care of personal matters during the day. This enables you to work in a way that enhances your productivity and also keeps you healthy.

Tip 129. Flowcharting

Flowcharting is the practice of creating diagrams that feature a "flow" of information and/or logic. The most popular flowcharting software is *Microsoft Visio*. See if your organization can provide you with this program. If not, then ask for permission to download its free and open-source alternative, *yEd* (https://www.yworks.com).

Tip 130: Forward, The Adverb

The adverb "forward" does *not* have an "e" between the letters "r" and "w". I look *forward* to completing the *Edge* admin guide. I will go *forward* with our plan to write Section Three of the new Tech Support guide.

Tip 131: Freemium (Factoid)

If you work and/or write as an entrepreneur, a popular revenue model to consider is called "freemium." This is where you offer a limited amount of written content for free and then charge a premium for the rest of the content. This is a "plausible" business model for many writers (including Tech Writers) across the board.

Tip 132: Functionality

Technical Writers learn about software functionality through requirements documentation and the input of SMEs and developers. Then, once understood, Technical Writers can discuss such functionality correctly in their user guides and training manuals.

Tip 133: General to Specific

For some systems, documentation is written from the general level down to specifics. Aircraft, for example, feature something called an "assembly." The greater assembly is composed of lesser assemblies. And these lesser assemblies can also have smaller subassemblies. This is what I call "system summary documentation."

Tip 134: Getting Needed Information

In my experience, information comes from both static and dynamic sources. The static sources are existing documents, websites, Intranet files, and corporate files. Dynamic sources are SMEs, other IT staff, managers, executives, and consultants.

Tip 135: Getting Needed Sleep

Please consult with your doctor to find out the best way for you to get steady, nightly sleep and rest. Thank you.

Tip 136: Getting Up to Speed

When you are new at a job, there is great pressure for you to learn the entire scope of what others already know. *Ask questions. Do not be shy.* Ask embarrassing questions. In this way, you will acquire important information quickly; soon you will be seen as a valuable company asset.

Tip 137: Glossary of Terms

When I write user guides and training manuals, I often include a "Glossary of Terms". I cannot define every technical term in the document at every step. So, I point the reader to the *Glossary of Terms* for a detailed explanation of each technical term mentioned.

Tip 138: Google Book

I wrote a book in 2014 called the *Google Productivity Guide* showing the entire scope of Google Platform apps and services. It is live at Amazon and still a great read. The Google Platform is still *amazing* and for that reason I wrote this book. For those of you on a shoestring budget, Google's free web/apps platform – Docs, Drive, Search, Play, Photos, and other programs - can really help you as a student and/or startup. The book's web address:

https://www.amazon.com/Google-Productivity-Guide-Charles-Johnson/dp/1502957736

Tip 139: Google Docs

For those of you who cannot afford Microsoft Office or who do not like LibreOffice, then I recommend *Google Docs*. This cloud-based office suite by Google is totally free and it features *Docs* for documents, *Sheets* for spreadsheets, and *Slides* for presentations. The user interface is extremely intuitive and the platform backs up your work immediately and frequently as you work.

I had one work experience where all documents were created/saved in Google Docs. Google Docs rocks!

Tip 140: Google Drive

For those of you who are Google enthusiasts, I highly recommend you use *Google Drive* to back up your work. Today, Google's Cloud Platform is more secure than ever. Technical Writers who write using different computers, devices, and/or locations will benefit greatly from the multi-presence and *trustworthiness* of *Google Drive*.

Tip 141: Google Search

There is no Internet Search Engine out there that can match the greatness of *Google Search*. In recent surveys, more than sixty percent of all core web searches (across all devices) are handled directly or indirectly using *Google Search*. The other almost-forty percent are handled by smaller engines like Bing (by Microsoft) and Yahoo. *Google Search is a very useful research tool for Tech Writers.*

Tip 142: Google Translate

Google Translate is a great program. It has both web and mobile versions. Essentially, you type in or speak the originating words in your selected language of origination and then select the output language. Click the visible "translate" button and then Google translates the input words into the output language. There are some Technical Writing jobs that require output documents in at least two languages. *Google Translate* is a great tool that can help ensure the accuracy of such deliverables.

Tip 143: Granularity

Granularity is a term used in business meaning *superb detail*. In the same way there is a grain or *granule* of sand or salt, information can be broken down so that we can see it in *fine* ways. Granularity allows conclusions about a system or program to be substantiated because we can, through such detail, understand how individual parts work at fundamental levels. Granularity is a sure path to solution-finding. *Technical Writers need to be able to break down processes and software aspects so that they can understand individual objects and see what happens to data at each step.*

Tip 144: Hacking

Hacking means to "use a computer to gain unauthorized access to a system." Hacking has become a major buzzword in the world of IT today. I stand by "ethical" hacking, which is a practice where IT staff test software/systems to see if access can be gained in ways that the system has been designed to prevent. *Technical Writers should become familiar with hacking concepts as they might have to cover this domain.*

Tip 145: Hardware vs. Software

Writing software documentation is not too hard. You write about the processing of data that users experience. Writing about hardware, however, is more challenging. You need to make sure that your technical details are spot-on accurate. There can be zero error when documenting a Disaster Recovery (DR) procedure, for example. Pay attention to detail. Work with OEM (Original Equipment Manufacturer) documentation when writing about hardware, if needed.

Tip 146: Health

The Buddha taught that "your health is your wealth".

Life is much better with health. Need I say more?

Tip 147: Highlighting, Screen Shot

Make sure that you highlight screen shots in accordance with your instructional text.

Example:

1. To visit Google's home page, type www.google.com in your browser's address bar, as follows:

2. Press *Enter* on your keyboard.
3. Navigate/use Google's home page as needed.

Tip 148: Homophone vs. Homonym

A *homophone* is a word that is pronounced the same as another word but is unique in meaning. A homophone might be spelled differently; the key to recognition is the similar pronunciation/sound.

Example: Write/Right. The developer will *write* the code snipped onto the white board. Navigate *right* on the next web page.

A *homonym* is generally spelled the same as another word but the two words have unique meanings.

Example: Bear/Bear. I can no longer *bear* the stress of this sprint; a *bear* crossed the road as I was driving into the office this morning.

Tip 149: HTML

HTML is an acronym that stands for *Hyper Text Markup Language*. HTML5 is the latest and most current version/standard. This is the primary "markup" language used to build websites.

Technical Writers should understand the basics of *HTML* so that they can support web developers.

Tip 150: Human Being then Techie

Technically-proficient entrepreneurs and executives know that we are all people first and then techies second. Highly prominent tech executives and entrepreneurs like Richard Branson and Elon Musk are known for being great humanists and they understanding the toils involved in taking on difficult tasks. They are not cruel. Rather they are helpful and supportive. They also know how to communicate technical information with care.

Practice compassion and empathy; you will then create great working relationships with those around you.

Tip 151: I Can (Mantra)

Repeat: *I can* complete this technical document.

Repeat: *I can* complete this user guide.

Repeat: *I can* coordinate SME and management input for this new user rights management guide.

Tip 152: Information Analysis

In some cases, Technical Writers will have enough information to be able to make certain observations and draw certain conclusions about functionality and process. In other circumstances, it is best that the Technical Writer consult with the SME responsible for said functionality or process. This way, your document is accurate.

I once joked with some colleagues that Technical Writing is like being a mad scientist at Jurassic Park. There, complete dinosaur DNA was created using 99% dinosaur DNA found in an ancient mosquito frozen in sap and 1% DNA from living frogs. Technical Writing is similar: about 99% of your technical content is verifiable and about 1%

of your technical info is not verifiable, so you need to "guesstimate" to the best of your ability. I have had to do this several times; my tech docs were approved.

Tip 153: Information and Obsolescence

In today's modern business world, systems that are implemented in many organizations are eventually replaced. I worked at a major electronic-payments software company for six years and documented three major system versions during that tenure. Each system attempted to work with new and emerging systems around it (for compatibility purposes). The fear of obsolescence propels new code. *New code propels new documentation.*

Tip 154: Information Gathering

There are *three* major human sources of information for my technical documents: Subject Matter Experts (SMEs), Business Managers, and Preferred Clients. The more "mindful system perspectives" that I obtain regarding a certain functionality, the more I am able to ascertain system design/intention. Technical Writers need to be perseverant regarding *documentation and validation.*

Tip 155: Information Readability

A manager once told me, "I cannot read this document." He gave me no tangible explanation. He simply "redlined" about half of the highly-detailed training manual I had finally written after much toil and research. Maybe this person did not have the patience needed to read this document. Was he having a bad day? Was he yelled at by his boss? *I do not know.* I have presented *official corporate documentation* to *top executives* on numerous occasions. Most of them, and most of the time, say my documents are understandable and well-written. I, therefore, wonder if a manager who tells me otherwise has a legitimate claim. Busy? Bad day? Fine. I do empathize. Still ...

Please do not shoot down a well-researched and well-presented document because you cannot find the time to read it or because you are having personal troubles.

Tip 156: Inheriting Work

Technical Writers, like other professionals, get assigned work under two general circumstances. First, you are the very first Technical Writer that your boss has hired, so there is little or no documentation written about the project(s) you are joining. Second, you are replacing someone else (1) who moved on, (2) who was dismissed, or (3) who quit. It is never easy to inherit work from someone else. In both scenarios, apply your *researching skills* to map out the entire system you are covering and know the related deliverables mapped out for your team. In this way, you will be able to hit the ground running.

Tip 157: Interviewing

Technical Writers need to be able to interview people. Recently, one product that I had to document featured RFID (Radio Frequency Identification) technology. This technology was being implemented so that inventory could be taken in more efficient ways. So, I prepared a list of questions to ask the software developer in an actual SME interview. The interview went well and the writeup also went well.

Having your interview questions prepared before an interview helps you to stay focused on the important and often highly technical content that must be understood and summarized in your notes and technical document.

Tip 158: Laptop

My favorite hardware resource as a Technical Writer is my laptop. With a *laptop computer* that has been set up properly, I can attend meetings in different conference rooms, be a road warrior and travel to other cities and countries, work from home, and still be totally productive. *Tech Writers can optimize work/time using a laptop.*

Tip 159: Libre Office

Libre Office is a free Office software suite by The Document Foundation that features just about everything you get with Microsoft Office – word processing, spread-sheet, presentation, and diagramming apps. If you cannot afford Microsoft Office, then get *Libre Office,* which is totally open-source and free of charge.

Tip 160: Life, Not Always Convenient

The business world can be chaotic. There are many reasons for this but accepting this fact is important if you are a Technical Writer. This will enable you to endure truly tough work days. You might be tasked with an important document that your boss needs by the end of the day, but your SMEs are not available or they are putting out fires with other matters and issues and so you get in line for their time. The clock ticks. Your boss is waiting. What can you do? In my humble opinion, there is only one solution: priority. If your boss tells your SMEs that this document is a top priority, then they can be *legitimately pulled away* from other matters, without being reprimanded by their bosses or losing their jobs. Then, you can meet with these SMEs and get your badly needed information.

Tip 161: LinkedIn

I have been active on the popular business networking platform called "LinkedIn" since 2008. I really like *LinkedIn*. You should too. If you are not on *LinkedIn*, you should join right away. Not only can you create a professional profile, but you can also microblog, write articles, network with others, and participate in communities related to your work. You can find job leads and get feedback from *LinkedIn* members regarding diverse professional issues and concerns. Tech Writers: Join *LinkedIn* if you can, it is really awesome.

Tip 162: Listening Effectively

Technical Writers obtain information from both static/written sources (online and onsite) as well as directly from dynamic/human sources like SMEs. When interviewing individuals with deep technical knowledge, you need to *listen effectively*. If you do not understand something, speak up. Otherwise, the SME will assume you have understood. Listen *with great attention* and focus in order to effectively understand. Listen oo that you understand well enough to be able to feature this new information accurately in your technical documentation.

Tip 163: Living Documentation

In the same way that software systems evolve, documentation evolves. This is why I usually include a Record of Revision page in my documents. So, as software is improved or functionality is revised, I include such information in my guides and such details are noted in the Record of Revision section.

Tip 164: Logging In, Logging Out

The term "log in" is a *verb* describing how you enter an app, website, portal, or platform that requires individual user credentials. The term "login" is an *adjective* used to describe credentials necessary for access (e.g. *login* credentials). Some Technical Writers prefer to use logon. I prefer login because you are *entering* a system. "Log out" and "logout" work the same way, respectively.

Examples:

Please *log in* to the portal using the web address given in this morning's e-mail. My *login* credentials are my corporate e-mail address for user ID and my last name followed by 123 for the password.

You can *log out* of the site by clicking on the *Bye* link in the top menu. The best *logout* method is to first save your work and then click on the *Bye* link.

Tip 165: Machine Learning

Some people think that AI and Machine Learning are the same thing. *They are close, but different.*

AI refers to highly-programmed computer systems that can perform complex tasks that we, as humans, consider "intelligent." *Machine Learning goes a step further.* Machine Learning is where we give a computer platform or system enough of a logical infrastructure at all levels so that it can start operating autonomously without further human support or input. Technical Writers should be aware of this subtle but definitive difference.

Tip 166: Management vs. Leadership

Technical Writers depend upon those in high places to pave the road ahead (e.g. managers/leaders securing resources that you will need in the coming days or weeks).

Tech Writers: Seek out those individuals who are great managers and great leaders. These are the very best people to serve; they will empower you and your path.

Tip 167: Managing Expectations

If a document request is such that it cannot be written by a mandated deadline, then speak up. You need to give the reasons why you cannot meet the deadline. *Document your progress.* Show your "work progress" spreadsheet periodically to your supervisor. Management can be aversive to documented facts, but if they really want the document, then they will be forced to listen and modify their short-term and long-term expectations.

Tip 168: Manuals

There are six major types of manuals/guides that Technical Writers normally compose:

User Manuals – these are written for both technical and non-technical readers and they can be about computer software, hardware, devices, apps, and more.

Tutorials – these are mini-guides that are akin to FAQ (Frequently Asked Questions) guides; they offer very limited content covering just specific/key topics.

Training Manuals – these are similar to user manuals but user manuals are usually much larger and cover more ground. Training manuals are written to help readers accomplish specific tasks. When writing a guide like this, have your supervisor *specify the tasks* to be covered.

Operations Manuals – you will find these types of manuals in (the dashboards of) cars, airplanes, and helicopters, for example; they are written to help the reader/user comprehend/interact with the product's UI/panel.

Service Manuals – many workers face processes that must be upheld so that the company can pass annual audits and reviews by official "bodies" like the FAA or FDA. In this way, companies stay in business. Workers must follow details in official *service manuals*, which show tasks to be done and how to do them, step by step.

Special-Purpose Manuals – if you are writing a manual that is not a user manual, tutorial, training manual, operator manual, or service manual, then you are composing a *special-purpose* manual.

Tip 169: Margins on Documents

The typical margin on documents I have written over the past two decades is one-inch on both the left and right side as well as at the top and bottom of the page. This amount of margin facilitates printing and PDF creation. If a document's text is too close to the sides and/or top or bottom of the page, there is a slight chance that your printer will not print all text. The same is true for PDF-generating software like Adobe Acrobat and Microsoft Office. You want readers to see all text and content in both hard copy and soft copy outputs. *Please choose your document margins wisely.*

Tip 170: Meditation, Technical Writing

Technical Writing can be very stressful. To offset any negativity you feel, close your eyes for one minute. Take a deep breath, hold it for three seconds, and release it slowly. This is a very simple meditation technique that can help you to release tension/stress.

Tip 171: Meetings

Meetings are helpful for all Tech Writers. Meetings are an opportunity to meet with people away from their desks. I have found in my career that people communicate better when they step away from their workstation(s).

Scheduled meetings are great because you can prepare for them. Impromptu meetings are helpful if you desperately need to meet with someone and they are very busy but tell you, "I am available now. Let's grab the conference room for fifteen minutes." Are you prepared for a spur-of-the moment meeting? Probably not. Still, do your best.

Tip 172: Microsoft Office 365 (and 2016)

Microsoft Office is the International Standard for Office Suite software used in businesses across the world. Microsoft became the great company it is today through not only DOS & Windows Operating Systems but also through Microsoft Office.

Microsoft Office features Word, Excel, Outlook, and other important software apps that helps people in offices across the world to create and manage documents. At the time of the writing of this book, Microsoft Office 365 (Microsoft Office 2016, for local installations) is the most current version of this cloud-based software suite.

Tip 173: Microsoft Excel Menus

Technical Writers should learn Microsoft Excel.

Excel is a spreadsheet program that is included with Microsoft Office 365 (Office 2016, for local installations). Spreadsheets are cell-based documents featuring tables of values arranged according to designated rows and columns. These rows and columns can be manipulated using the tools and resources that Excel provides. Sorting, simple calculations, and advanced calculations are common user activities.

Major Excel Menus and Menu-Options:

File: Info, New, Open, Save, Print, Share, Export, Publish, Close

Home: Clipboard, Font, Alignment, Number, Styles, Cells, Edit

Insert: Pivot Tables, Illustrations, Add-Ins, Charts, Symbols

Page Layout: Themes, Page Setup, Scale to Fit, Sheet Options

Formulas: Function Library, Defined Names, Formula Auditing, Calculations

Data: Get/Transform Data, Queries/Connections, Sort/Filter, Tools

Review: Proofing, Accessibility, Language, Comments, Protect, Ink

View: Workbook Views, Show, Zoom, Window, Macros

Help: Help, Contact Microsoft Support, Feedback, Community

Tip 174: Microsoft Excel Basic Ops

Creating a Spreadsheet

First, go to File > New. Then, select a new document or one of the visible document templates (by double-clicking on the image/icon). Your new spreadsheet opens.

Saving a Spreadsheet

First, make sure your document is loaded and live within Excel. Then, go to File > Save for current documents or Save As for new documents. Make sure you save using a meaningful name. Also verify the document extension just to be sure/safe. Finally, make sure you choose a location on your local drive or the cloud that is easily accessible for continued work.

Adding Sheets

You are not limited to the first sheet that your spreadsheet presents. You can add additional sheets. At the bottom of your Excel spreadsheet, you will see the current tab (which should be highlighted in white). Click on the plus (+) icon that you see and a new sheet is created, immediately. If you right click on the sheet, you can create a name that makes sense. You are not obliged to stay with Sheet1, Sheet2, etc.

Getting Data

You can use Excel to get and analyze data. In some cases, you have existing Excel spreadsheets that just need to be opened. If they were written in a previous version, Excel will inform you and ask if you want the document upgraded to the current version. Choose Yes unless you are returning the document to someone who does not have your version of Excel. When data comes from another location or from another format, go to the Data tab and use the Get Data option. This option enables the importing of data into your current version of Excel. Now, you can review what you have imported.

Moving Information to other Office apps like Word

Select, using your mouse, the Excel cells and/or diagrams you need for your Word or PowerPoint document. Press Ctrl-C on your keyboard to copy. Go to your Microsoft Word document, for example. Place your cursor; click Ctrl-V to paste. You have copied/moved the content.

Exporting to CSV

The acronym CSV stands for *Comma Separated Values*. This is a popular and universal file format for excel and other types of data files.

Go to File > Export > Change File Type. You will see a pop-up window. Verify the file's name. Select CSV for file format. Click Save. The document is now in CSV format.

Tip 175: Microsoft Excel, Keyboard Shortcuts

Close a Workbook: Ctrl + W

Open a Workbook: Ctrl + O

Go to Home Tab: Alt + H

Save a Workbook: Ctrl + S

Copy: Ctrl + C; Paste: Ctrl + V

Cut: Ctrl + X; Undo: Ctrl + Z

Remove Selected Cell Content: Ctrl + Z

Bold: Ctrl + B

Go to Page Layout Tab: Alt + P

Go to Data Tab: Alt + A

Go to View Tab: Alt + W

Tip 176: Microsoft Excel Help

Go to: https://support.office.com/en-us/excel

Tip 177: Microsoft Outlook Menus

Technical Writers should learn Microsoft Outlook.

Outlook is an electronic communications software for email and meetings that is included with Microsoft Office 365 (Office 2016, for local installations). Outlook can also be called a personal information manager from the general-utility perspective. This software application features an email client, calendar, contact list, task manager, activity journal and more.

Major Outlook Menus and Menu-Options:

File: Info, Open/Export, Save (Attachments), Print, Support, Options, Exit

Home: New, Delete, Respond, Quick Steps, Move, Tags, Groups, Find

Send/Receive: Send/Receive, Download, Server, Preferences

Folder: New, Actions, Clean Up, Favorites, Online View, Properties

View: Current View, Messages, Arrangement, Layout, People Pane, Window

Help: Help, Contact Support, Suggest Feature, Show Training

Tip 178: Microsoft Outlook Basic Ops

Open, Receive Email

First, click on the appropriate folder in your email client. This usually is in the left column or left pane. Then, select the email to be opened by double-clicking on it with your mouse. From your mobile device, you most likely will only have to single click on the email message to open it. The email message will display fully in the Content pane. You will see its origin, content, signature, and attachments, depending upon what the sender included.

Compose, Send Email

In Outlook, find and then click on Home > New Email. A new Email message window displays. Enter recipients, subject, contents, and review. Then, click on the Send button to send it out. It will first display in your Outgoing folder. Then, you will see the message listed in your Email's Outbox folder.

Attaching a Document to your Outgoing Email

First, make sure you have composed all or most of your outgoing email message in the New Email Window. Then, go to Insert > Attach File. A pop-up window displays. Navigate to the folder in which your file to be selected is stored. Go to this file and double click on it. Outlook attaches this file to your email message. Note that not all file formats can be attached, especially executable files, because sometimes they can have a virus. But, you should be safe with Word docs, Excel docs, text docs, PowerPoint docs, and PDFs. When all is ready with your message and attachment(s), click on the Send button to send it out.

Locating your Calendar Menu

At the bottom of your Outlook email client are normally tabs for email and calendar. Click on the icon that looks like a Calendar. In Calendar view, you can schedule meetings and use the tools and resources of this aspect of Outlook to find meeting rooms, invite participants, include meeting details and more.

Tip 179: Microsoft Outlook, Keyboard Shortcuts

Close: Escape or Enter key; Send: Alt + S

Go to Home Tab: Alt + H

New Message: Ctrl + Shift + M

New Task: Ctrl + Shift + K

Delete: Delete key (but first select item)

Go to Calendar: Ctrl + 2

Go to Mail View: Ctrl + 1

Go to Contacts View: Ctrl + 3

Tip 180: Microsoft Outlook Help

Go to: https://support.office.com/en-us/outlook

Tip 181: Microsoft Planner Overview

Technical Writers should learn Microsoft Planner.

Planner is a completely new app for Office 365 and it does not have a local version for Office 2016 (as of yet). It is web-based only (for right now). Planner is essentially a dynamic online bulletin board where tasks are created through online "cards".

Managers can set up how employees have their work flows and cards can be dragged and dropped to different stages of a work flow. As a case in point, you can have Planner Categories or "Buckets" such as Ideas, Works in Progress, and Complete. Managers can establish new tasks, using the Planner cards, and task workers accordingly. Depending upon rights given, managers and/or workers can move the cards from Ideas, to Works in Progress, to Complete. Planner cards have sections where you can create a title, description, and include attachments.

Notifications through email are possible so when documents are due, workers get notified just before due dates. Planner is a very cool app.

Tip 182: Microsoft Planner Help

Go to:

https://support.office.com/en-us/article/microsoft-planner-help-4a9a13c6-3adf-4a60-a6fc-15c0b15e16fc

Tip 183: Microsoft PowerPoint Menus

Technical Writers should learn Microsoft PowerPoint.

PowerPoint (PPT) is a great presentation software developed by Microsoft to help and empower business and social communications with visual aid resources. PPT has both simple and sophisticated tools that enable simple one or two slide presentations and also enable long multi-slide presentations that feature rich-multimedia like layered text, audio, and video inclusion.

Major PPT Menus and Menu-Options:

File: Info, New, Open, Save, Save As, Print, Share, Export, Close

Home: Clipboard, Slides, Font, Paragraph, Drawing, Editing

Insert: Slides, Table, Images, Illustrations, Links/Comments, Text/Symbols, Media

Design: Themes, Variants, Customize, Designer

Transitions: Preview, Transitions (to this slide), Timing

Animations: Preview, Animation, Advanced Animation, Timing

Slide Show: Start Slide Show, Setup, Monitors

Review: Proofing, Accessibility, Insights, Language, Comments, Compare, Ink

View: Presentation Views, Master Views, Show, Color/Gray, Window, Macro

Add-Ins: Custom Toolbar

Help: Help, Contact Support, Show Training

Tip 184: Microsoft PPT Basic Ops

Create Presentation

Open PowerPoint (PPT) and then go to File > New. Select either a new/blank presentation or a template that Microsoft offers. Double-click to open. Create your presentation.

Save Presentation

Go to File > Save (existing presentations). Go to File > Save As (new presentations or current presentation to be saved under a new or different name).

Tip 185: Microsoft PPT, Keyboard Shortcuts

Create New Presentation: Ctrl + N

Make Selected Text Bold: Ctrl + B

Cut selected text and/or object(s): Ctrl + X

Copy selected text and/or object(s): Ctrl + C

Paste selected text and/or object(s): Ctrl + V

Undo last action: Ctrl + Z

Save Presentation: Ctrl + S

Tip 186: Microsoft PPT Help

Go to: https://support.office.com/en-us/powerpoint

Tip 187: Microsoft SharePoint Overview

Technical Writers should learn Microsoft SharePoint.

SharePoint is a cloud-based file-management platform that is included with Microsoft Office 365 Professional. SharePoint is a powerful platform that is used to create and manage intranets, websites, document repositories, wikis, blogs, and more.

Tip 188: Microsoft SharePoint Help

Go to: https://support.office.com/en-us/sharepoint

Tip 189: Microsoft Skype Overview

Technical Writers should learn how to use Skype.

Microsoft bought *Skype* in 2011 and includes it today with Microsoft Office. *Skype* has both cloud and local app installations/access. *Skype* is a VoIP platform meaning *Voice over Internet Protocol*. So, both text and video communications are done through the Internet itself as opposed to using a traditional communications carrier. *Skype* is extremely user-friendly. You can communicate directly with your contacts through *Skype IM* (Instant Messenger) for text-only purposes as well as use the platform to communicate with them using audio and video. Make sure *Skype* has access to your device's camera and audio as well for best results/use.

Tip 190: Microsoft Skype Help

Go to: https://support.office.com/en-us/skype-for-business

Tip 191: Microsoft Visio Overview

Technical Writers should master Microsoft Visio.

Visio for Office 365 Professional (and for Office 2016 Professional) is an app that requires paid subscription. Visio is a powerful diagramming tool and resource that will help you to create designs, diagrams, flowcharts and more to support your technical text and content.

Tip 192: Microsoft Visio Help

Go to: https://support.office.com/en-us/visio

Tip 193: Microsoft Word Menus

Technical Writers should master Microsoft Word.

Microsoft Word is a graphical word processing application that Technical Communicators – especially Technical Writers – use to create and manage documents. I remember when Word 1.0 was released for my Mac Plus computer while I was in college in the late 80s. Today, we have Word 365 in the cloud and Word 2016 for local (e.g. local hard drive) installations. Word is versatile. It has amazing features useful to Technical Writers.

Major Word Menus and Menu-Options:

File: New, Open, Save, Print, Share, Export, Close

Home: Clipboard, Font, Paragraph, Styles, Editing

Insert: Pages, Tables, Illustrations, Headers, Footers

Design: Document Themes, Document Formatting, Page Background

Layout: Page Setup, Paragraph, Arrange Text, Arrange Objects

References: Table of Contents, Footnotes, Research, Citations, Captions, Index

Mailings: Create, Start Merge, Write/Insert Fields, Preview Results

Review: Proofing, Comments, Tracking/Changes, Protect

View: Views, Show, Zoom, Window, Macros, SharePoint

Help: Help, Contact Microsoft Office Support, Feedback

Tip 194: Microsoft Word Basic Ops

Creating a Word Document

Open Microsoft Word and go to File > New. Select a blank document template or a formatted template that Microsoft displays. Double click on the icon for the type of document you are going to create. Continue ahead and compose your document.

Saving a Word Document

While you are in Microsoft Word, you will be working on either a brand-new document or an existing document. For brand-new documents, go to File > Save As. In the pop-up window that displays, enter the file's name, file location where it will be saved, and verify the file format. Click Save. For existing documents, you will not need to go through this pop-up window unless you are changing the document's name or location. Go to File > Save or press Ctrl-S on your keyboard to save the latest and greatest version of your document.

Editing a Word Document

While you are in Microsoft Word with an active document, use your cursor via your mouse or mousepad to select blocks of text and then make updates. Also, place the cursor where it is needed and either enter or delete text and/or graphics as needed. If needed, you can select Ctrl-A on your keyboard to select all text and graphics in the document. Use the View > Zoom option in your Main Menu to see the document in different sizes to ensure its presentation meets your expectations and goals.

Microsoft Word Headers and Footers

Double click in either the header or footer region of your document. The display of the page changes such that you can now access these areas. Insert or delete what you must. You can, for example, put the document's name in the Header. You can, for example, paginate in the footer. Double click in the body of the document to get out of this mode and return to regular content mode.

Microsoft Word Document Orientation

The two most common document orientation choices are Portrait and Landscape. Portrait is "upright" and landscape is "horizontal", if you will. Choose as you need. Go to Insert > Orientation > Portrait or Landscape. Use Ctrl-S to save updates to document orientation.

Microsoft Word's Best Document Fonts (IMHO)

The best fonts to use for professional documents are *Calibri, Palatino,* and *Verdana.* All of these fonts render well universally on screens and on paper after printed. They are clear and indirectly help readers to make their way through long and complex documents with minimal eye strain. My "top font" of all times: *Calibri.*

Tip 195: Microsoft Word, Keyboard Shortcuts

Open Doc: Ctrl + O; Select Entire Document: Ctrl + A

Cut selected Text and/or Objects: Ctrl + X

Copy selected Text and/or Objects: Ctrl + C

Paste selected Text and/or Objects: Ctrl + V

Find specific Text: Ctrl + F

Tip 196: Microsoft Word Help

Go to: https://support.office.com/en-us/word

Note: The following two tips, alphanumerically, should be listed before Microsoft Word. However, I wanted to keep *continuity of concept* while presenting Microsoft Office. So, I have placed tips for Windows 10 & Windows Movie Maker on this page (e.g. the next two tips).

Tip 197: Microsoft Windows 10

While I respect Linux enthusiasts and Apple enthusiasts, I stand by personal computers that run the Microsoft Windows Operating System (OS). Back in the 1990s, while living in Brazil, I developed two desktop applications for very early versions of Microsoft Windows and have stuck with the Windows OS for the past twenty-plus years. Today, Windows 10 is not only the latest but also the greatest OS that Microsoft has released.

It features a truly intuitive user interface and the apps included are all very useful. Windows has always been supportive of all kinds of amazing applications over the years. In 2016, I wrote two books about Windows 10 to help users maximize their W10 experience: *Windows 10 Productivity Guide*, and *Windows 10 for Seniors and Beginners*. Both are live at Amazon.com.

Tip 198: Microsoft Windows Movie Maker

If you need a free video creating and editing resource, you can use *Microsoft Windows Movie Maker*. It is no longer included with Windows 10 but there are sites on the web like www.en.softonic.com that offer free downloads of this great multimedia tool. WMM is super-easy to learn.

Tip 199: Microsoft Writing Style Guide

My favorite Technical Writing reference/resource is the *Microsoft Writing Style Guide*. It details many topics and terms that Technical Writers cover. It web page is:

https://docs.microsoft.com/en-us/style-guide/welcome

Tip 200: Microsoft Visual Studio IDE

The Microsoft Visual Studio Integrated Development Environment (IDE) is awesome. I urge all Technical Writers to learn the basics of this tremendous software development platform, especially if you are going to document code written here. Developers can use the Visual Studio IDE to write code for many different ends: desktop apps, mobile apps, and web apps. Visual Studio features *IntelliSense*, which is instant context-aware help designed to help the developer make correct choices and type things in correctly in the provided text editor. This tool helps programmers to expedite final code.

The built-in documentation tool for Visual Studio is called *Sandcastle.* Tech Writers might be asked to use *Sandcastle* to document code written in this IDE.

Tip 201: Mindful Reader Feedback

As a Technical Writer, I find great value in mindful reader feedback. I appreciate when both technical details and presentation are reviewed from a professional point of view. It feels great when people tell me "this document looks good" but it is better when someone says "OK, this document is good, but verify with SME ABC to ensure that your facts on page ZZ are accurate."

Tip 202: Mobile Devices (Touch Gestures)

If you are writing a user guide or a FSD (Functional Specifications Document) for a mobile app, you will face the UI (user interface) common to mobile devices: *touch gestures*. There are many common touch gestures between Android and iPhone Operating System interfaces. There are also many unique gestures that you should know and understand too. Here is a free online guide that nicely summarizes *touch gestures* for mobile devices:

https://www.lukew.com/ff/entry.asp?1071

Tip 203: My Favorite Open Source Software

This tip is for students/individuals with a low budget.

Note: The OS tip is for Windows PCs only. The apps are device-agnostic.

Operating System (OS): Ubuntu – free OS.

Office Suite: Libre Office – free app.

Malware Protection: Advanced System Care – free.

Diagramming Software: yEd – free app.

Graphics Software: Gimp – free app.

Web Browser: Mozilla Firefox – free app.

Tip 204: My Favorite Proprietary Software

This tip is for working professionals.

Note: The OS tip is for Windows PCs only. The apps are device-agnostic.

Operating System: Microsoft Windows 10.

Office Suite: Microsoft Office 365.

Malware Protection: Advanced System Care (there are both free and paid versions).

Diagramming Software: Microsoft Visio.

Graphics Software: Adobe Photoshop.

Web Browser: Google Chrome.

Tip 205: My Favorite Time

Early morning is my favorite time of the day.

I know that many writers are night owls and live by strange hours. I understand. I used to be a night owl myself. Currently, however, I like to get up early at daybreak, take a brisk shower, put on fresh clothes, and enjoy a healthy breakfast and espresso coffee. Then, when I sit down at my computer workstation, I am fully energized and ready to go. The mystics of ancient India call early morning hours "sattva guna", in Sanskrit, meaning "good energies". So true!

Tip 206: My Favorite Zen Saying

"Sit still" – Stillness yields mental clarity.

Tip 207: My Favorite Question: "Why?"

Tech Writing involves the documenting of functionality.

The Tech Writer who shows the reader both (1) how to accomplish a certain task and (2) why that task is important is the greatest of Tech Writers, in my book.

Tip 208: Neither, Nor

When writing to address "lack," a great literary mechanism is the use of two terms: "neither" and "nor." *Do not use one without using the other.* They go together as a pair. Here are three examples:

- *Neither* the CEO *nor* the CTO attended the executive meeting this morning.
- I will be researching the web using Microsoft Edge as my browser. I will be using *neither* Google Chrome *nor* Mozilla Firefox because our company mandates the use of Microsoft resources for all work activities.
- *Neither* the Production databases *nor* the Reports databases are accessible right now.

Tip 209: Notepad++

Notepad++ is an awesome and dynamic text editor and source code editor. Notepad++ supports tabbed editing, which allows one to work with multiple open files in a single window. So, for all you Technical Writers out there looking for a truly great (and free) technical text editor as you work to support developers, try Notepad++.

Tip 210: Notes

One key activity of Tech Writers is note-taking.

In some instances, you will not have to write many notes because the meeting organizer has pre-written information that you just need to follow. In other instances, you will be learning about sophisticated systems and/or processes and you will have absolutely nothing prepared by the meeting organizer. What can you do? Well there are several possibilities.

First, you can write as you listen. You can also type as you listen. That is only partially efficient. Second, you can ask the meeting host if you can record the meeting using a program like Audacity. Third, you can also ask that most of the meeting be held using a whiteboard and, using a company-authorized camera, take pictures of the final diagrams created on the white board. Remember to coordinate your notetaking in meetings with SMEs, co-workers, and managers to make sure that your tools and approaches are accepted by your organization.

Tip 211: Nouns

The two most important parts of speech in the English language are nouns and verbs. Sure, adjectives and adverbs matter. But, without nouns and verbs, the English language loses its essential base. Nouns and verbs, together, establish language presence. I will cover verbs toward the end of this book.

There are two types of nouns: proper and common.

Proper nouns are used for specific people, places and things. Proper nouns are always capitalized. Examples: Bill Gates, Steve Jobs, Jeff Bezos, Miami, New York City, Paris, London, Dell, Amazon, Microsoft, Google, Intel, Latin America, Europe, and Asia.

Common nouns have the following sub-categories: singular, collective, concrete, abstract, count, and mass. Singular nouns imply one unit only. Collective nouns imply a group of single units. Concrete nouns are about tangible objects or entities. Abstract nouns are about non-tangible entities. Count nouns are those that can be quantified. Mass nouns are those that cannot be separated into countable units.

Here are some examples:

Singular Nouns: bus, airplane, person, citizen, employee, workstation, device.

Collective Nouns: jury, ensemble, pack, department, band, team, committee.

Concrete Nouns: paper, computer, printer, cable, television, car, pen, tie, suit.

Abstract Nouns: process, effort, meeting, decision, incentive, title, reason.

Count Nouns: desks, printers, calculators, pencils. pens. books. envelopes.

Mass Nouns: electricity, energy, air, water, fuel, pressure, temperature.

Tip 212: Objects

Objects are nouns that are the "focal point" of transitive verbs. Intransitive verbs lead usually to prepositional phrases while transitive verbs have objects. Please note that I will cover and address verbs toward the end of this book. Just look up "verbs" in the Table of Contents to find the "Verbs" tip. Here are a few examples of objects:

The technical writer *composed* a user guide.

The mathematician *wrote* the formula on the whiteboard.

The report writer *produced* the year-end-summary for management.

The baker *created* a cake for the customer's wedding.

The dog *buried* the bone in the back yard.

The student *passed* the written examination.

Tip 213: Online Communities

I am a believer in *Online Communities*. There are *Google Groups*. There are *Yahoo Groups*. There are also Professional Groups at LinkedIn. Get involved. Subscribe. Participate. Offer information. Receive information and feedback. We are a collective race (e.g. humanity) and we all benefit the more we interact. Grow/evolve with an online community!

Tip 214: Organizational Charts

Over my twenty-plus year career, I have composed many different organizational charts. Some charts are only to show key relationships among senior management. Other charts are designed to show all personnel of an organization from top to bottom.

Organizational charts are helpful because when you are seeking specific information, you can find the department as well as managers and staff who will become your source. You can use PowerPoint and Visio to create effective and informative organizational charts.

Tip 215: Outlines

Creating an outline is helpful but only after you brainstorm and finalize your "information inventory" and "information scope". Then, you can create an outline that spans the distance and depth of your topic to be covered. Here is the general outline of my recent book "The Windows 10 Productivity Guide," as an example:

1. Introduction
2. Downloading/Installing Windows 10
3. The Windows 10 User Interface
   a. Start Menu
   b. Metro Windows
   c. Task Bar
   d. Desktop

4. Windows 10 Apps
5. Windows 10 Settings
6. Summary
7. Reference Section

Tip 216: Pagination

Pagination is the act of assigning specific numbers to the pages of your document. Here is my personal and also professional take on pagination: Any document that is up to nine pages in total can go without pagination. You should apply pagination only to documents that are ten pages or more in length (including the cover page).

Tip 217: Paragraphs

I like to call paragraphs the "beat" of a document. In the same way that a song has a certain "beat," a document also "moves along" at a certain pace. There are some songs that are fast and superficial and there are some that are slow and deep. The same holds true for documents. Some are fast for reading; paragraphs have two or three sentences and the content is superficial. Others are slow for reading; paragraphs have six or seven sentences and the content is rich and complex. Paragraphs serve three purposes. First, they *initiate* an idea. Second, they *carry* an idea. Third, they offer the writer a chance to *control* how the reader consumes the information presented.

Example of a superficial paragraph:

The programmer wrote remedial code all night. He compiled the code in the early morning hours. When his boss arrived at noon, the system was operational.

Example of a deep and rich paragraph:

The technical writer was tasked with documenting the programmer's source code. First, he downloaded a free and open-source code documentation tool off the web. Second, he copied the programmer's files from the production platform into a folder on the testing platform. Third, he started the code documentation tool, specifying the programming language to document and the location of the files. Fourth, he clicked Run. After waiting thirty minutes, the software informed him that it successfully documented the source code files.

Tip 218: Parallelism

Parallelism is the practice of presenting information in a consistent way, from a grammatical standpoint. Parallelism is especially useful in technical documents because it increases readability. As you establish your style, the reader will soon anticipate upcoming paragraphs, lists, etc. With that expectation met, the reader will gain confidence and be able to consume the information with less mental strain.

Instructions without Parallelism

1. Go to Google's home page.
2. The URL should read www.google.com
3. Now you can enter your Search topic.

Instructions with Parallelism

1. *Go* to Google's home page.
2. *Enter* www.google.com in your web browser's address bar.
3. *Type* in your topic into the Search box.

Tip 219: Paraphrasing

Paraphrasing is the practice of taking what someone has said or written and resaying it or rewriting it in a way that is unique in words and unique in presentation. If you rewrite exactly what someone has said or written, it must be presented in double quotations and sourced.

Tip 220: Passive Voice

Passive voice is the literary opposite of active voice. As a general best practice, active voice is to be used as much as possible.

According to my literary sources, passive voice is appropriate *when the receiver of an action needs to be emphasized more than the giver of an action.*

Examples:

The database update is performed by the DBA each night at six o'clock.

Note: The database update is more important than the DBA.

The operations manual was written by the quality analyst.

Note: The operations manual (OM) is more important than the quality analyst (QA).

Tip 221: Parts of Speech

All words in the English language are a specific "part of speech." This is a term used to describe basic language components of the English language. The result of combining "parts of speech" correctly is meaning.

The English Language's major parts of speech are:

1. Noun – mechanism used to name something or someone
2. Pronoun – substitute for a noun or noun-phrase
3. Verb – mechanism used to express action
4. Adjective – mechanism used to describe nouns and pronouns
5. Adverb – mechanism used to describe verbs and adjectives
6. Conjunction – mechanism joining two sentences

7. Preposition – mechanism that shed lights on verbs in light of nouns
8. Interjection – mechanism that expresses an exclamation

Examples:

1. Nouns: computer, software, hardware, process, program, application
2. Pronouns: he/his, she/her, it/its, they/their, you/your
3. Verbs: start, process, complete, review, submit, enhance, describe
4. Adjectives: strong, intuitive, clear, vague, insightful, brilliant, helpful
5. Adverbs: very, incredibly, slightly, nearly
6. Conjunctions: and, if, but
7. Prepositions: phrases beginning with words like in, by, for, with, etc.
8. Interjections: Hey! Yo! No way! Stop! Go!

Tip 222: Path

In today's highly interwoven business/tech worlds, the term "path" is often used across the board to refer to the way you get to a platform, environment, server, intranet or internet site, document, folder or archive.

In my experience, you can use "path" when referring to a static (non-changing) point of reference.

For example: The *path* to my new document is C:\Documents\Software.

You can also use "path" when stating navigational steps.

For example: Here is the navigational *path* to my new document: My Computer > Documents > Software.

Tip 223: PDFs

PDF is an acronym that stands for *Portable Document Format*. PDFs are able to open in just about any Operating System: Linux, Windows, Apple OS, Android, etc. PDFs can also be opened with the assistance of your web browser. If you use Microsoft Word or LibreOffice Writer, for example, you can always save secondary copies of your documents as PDFs; I do this infrequently.

If you need a tool that can enable you to join or separate individual PDFs, then I recommend *PDF Sam*. I have used this app a few times; it works really well.

Tip 224: Peer Editors

In the business world, a "peer" is your co-worker. It can also be visiting contractors or consultants, managers, and even the owner of a company. Peer does not mean totally lateral status within an organization.

For example, if you are a Junior Technical Writer, it is not just other Junior Technical Writers or Junior Associates who are your peers. Rather, your peers can be your supervisor, fellow employees, contractors, and SMEs. Peer editors are a powerful means of *ensuring total accuracy* in your document. They see things from their vantage point.

Tip 225: Pen and Paper

In today's highly technical and mobile world, working with a simple pen and pad of paper seems to be a thing of the past. But, this is not true. There is great power and grace in writing with a simple pen (on paper). A pen is a simple writing instrument you hold with your hand. The pad or piece of paper is something upon which you write. You do not have to worry about saving a file. Just write!

Tip 226: Per

The word "per" is overused today both in and out of the office. According to my sources, "per" is most appropriately used for technical references. It should not be used in daily dialogue.

Incorrect Use

Per your suggestion, I will back up my work on a daily basis.

Correct Use

You can have one user assigned to this station *per* paid login credential.

Tip 227: Percent, Percentage

Use the word "percent" as opposed to its symbol "%". However, you can use the symbol in tables like those in spreadsheets. Use the word "percentage" in the same way and manner.

Incorrect Use

The hard drive is 50% full.

Correct Use

The hard drive is 50 percent full.

Tip 228: Period

The period is called a "punctuation mark" of the English language. Periods officially end sentences. I highly recommend the use of periods in technical documents over semi-colons. The period means this: Stop Reading. Process Information. Continue Reading.

Tip 229: Permissions

The word "permissions" should apply exclusively to operations regarding shared resources (e.g. files, directories, printers, and web pages like portals). Permissions are normally "assigned" to end users by empowered system administrators. Remember that permissions are "granted" or "assigned" and not "allowed" or "permitted."

Tip 230: Person

Person refers to the "instance" of an individual or individuals in a sentence. The three "instances" of person, in the English language, have both singular and plural forms. Please take notice, Tech Writers!

They are as follows:

First Person Singular Forms (I, me, my, mine)

First Person Plural Forms (we, us, our, ours)

Second Person Singular Forms (you, your, yours)

Second Person Plural Forms (you, your, yours)

Third Person Singular Forms (he, him, his, she, her, hers, it, its)

Third Person Plural Forms (they, them, their, theirs)

Tip 231: Phenomenon (Factoid)

Something that is observable is considered to be a "phenomenon". The word "phenomena" is its plural form.

Tip 232: Photos

There are many different technical manuals that can benefit from the inclusion of actual photographs. What you need to do is coordinate the "photo-inclusion process" with your supervisor and also auditing authorities. This way, such photos are formally accepted during audits. I worked for two years at a manufacturing company in Miami and wrote CMMs (Component Maintenance Manuals). I was allowed to include actual photos of safety product components to shed light on technical text.

Note: If you use a photograph taken by a professional photographer, make sure you do two things: (1) obtain written permission from the photographer to use his/her work in your document and (2) attribute the photo to its source correctly and completely.

Tip 233: Physical Books vs. Electronic Books

I was a high-school student and college student in the 1980s. At that time, we did not have *either* the Internet *or* mobile phones. We had to use physical books. Today, students and working professionals have both. I still love the feel of a physical book. You can write in it. You can underline and/or highlight text. You can place comments in empty spaces in the book. This close interaction helped me to learn and grow as a student.

Today, this interaction still helps me to continue to grow and evolve. Electronic books – whether in PDF or Kindle or Nook format – are convenient for sure. Still, they lack the tangibility that some people enjoy as they read. The choice is yours, but please do consider enjoying a physical book as a way of periodically breaking from your mobile device apps, calls, texts, and notifications.

Tip 234: Phrases

A phrase is a group of words that have meaning but they do not, by themselves, constitute a complete sentence. There are six types of phrases that you can use to improve your technical documents.

(1) Prepositional phrases
(2) Participial phrases

(3) Infinitive phrases
(4) Gerund phrases
(5) Verb phrases
(6) Noun phrases

1. <u>Prepositional phrases</u> feature a preposition, an object, and an object modifier.

<u>Example</u>: The meeting started without my managerial authorization.

2. <u>Participial phrases</u> feature a participle, an object, and an object modifier.

<u>Example</u>: The company, having the best sales software, will broadcast its town hall meeting live on Facebook tomorrow.

3. <u>Infinitive phrases</u> feature a verb starting with "to".

<u>Example</u>: Google tries to provide an unparalleled Internet search engine.

4. <u>Gerund phrases</u> feature a gerund plus object and modifier.

<u>Example</u>: Training employees is a time-consuming task.

5. <u>Verb phrases</u> feature two verbs: a main verb and a "helper" verb.

<u>Example</u>: The Facebook intern is learning the Python programming language.

6. <u>Noun phrases</u> feature a noun and a modifier.

<u>Example</u>: Let these three experienced security experts handle this matter.

<u>Tip 235: Plagiarism</u>

Unless your *choice of words* is one-hundred percent original, then you must cite the information's source properly at the end of your document.

The *Chicago Manual of Style* will show you how to properly cite such information in your work.

Tip 236: Planning

Planning is very important for Tech Writers. This skill enables you to map out a document, for example, *as a project*, and establish *milestone*s and *steps* you will take to accomplish each one. Planning gives you positive leverage because you can always step away from the writing and look at the big picture. For example, I am elaborating on this writing tip, but I can always revert back to the document's TOC to see where I stand as I write this book, *365 Technical Writing Tips*. Perspective!

Tip 237: Platform

The technical term "platform" refers to "computing environment," usually, in the world of Information Technology. The emerging "platform" of modern times is "the cloud," which refers to a connected and interactive network of remote servers hosted both applications and data. Microsoft Azure and Google Cloud are both well-known and high-quality cloud platforms. Many Tech Writing jobs/gigs today involve the documentation of platforms, like Software as a Service (SaaS).

Tip 238: Presentations

Presentations have three parts: *Introduction, Main Content,* and *Summary.* Whether you are creating a PPT presentation or a training guide, remember what I have stated here. The more complex the training, the more you are challenging the participants and their ability to follow. *So, stay on track with these three simple parts/sections.* Participants will remain focused and thus benefit from your presentation/training session.

Tip 239: Prioritization

When I begin a Technical Writing gig, I am often tasked with a "starter" document. It could be a simple user guide. It could be a simple training manual. Shortly thereafter, my supervisor will task me with tougher documents, like writing Admin guides. Within a month, I will have probably at least ten major documents on my plate. How do I manage them? *The answer is prioritization.*

What I do is sit down with my supervisor or manager and itemize all the documents I am expected to write along with their projected due dates. This up-front manner of *prioritizing documents* right there with one's boss is a great way to work: both parties are on the same page.

Tip 240: Professionalism via Clarity

Writers need to *compose clear sentences.* Words need to be selected carefully. You also need to make sure that your reader did not get lost a few pages ago. Clarity is the result of both (1) selecting quality words and (2) applying presentation best practices.

Tip 241: Professionalism via Compassion

Compassion is another great word for empathy. Compassion does not mean that you feel sorry for your reader. Rather, you are aware of the personal and professional challenges that your reader is presently managing. Hence, you write with great care, detail, and surety. Your goal is to *help the reader,* slowly and steadily, to master the technical content you are presenting.

Tip 242: Professionalism via Curiosity

Tradition has it that cats are always curious. Well, dogs are curious too. Technical Writers need to put fear aside and be curious like cats and dogs. Why does the button-click only show this? Why does it not show more?

Why does the Menu only feature four options when it could really feature ten? Technical Writers sometimes have to venture into Quality Assurance (QA) and testing grounds as they write.

Curiosity is a *positive means to discovery,* especially when you work with few or no system requirements. Curiosity is what allows you to discover what the developer has coded/created. In this way, your tech doc will shine!

Tip 243: Professionalism via Generosity

Generosity is what drives me *every single day* as a Technical Writer. I sincerely care about my reader's success. The spirit of generosity helps us to see that our work is not in vain. Rather, we are generating a huge benefit to others through a well-written technical document, user guide, or manual. So, this is why the work of Technical Writers is very important. We create user guides and technical documents that help new employees and that address unique technical topics that readers need to understand in order to do their jobs well.

The user guide you are writing, for example, is not a meaningless document. Rather, you are applying your *spirit of generosity and know-how* through the document and are positively empowering others. Good job!

Tip 244: Professionalism via Patience

There is a common saying: "patience is a virtue." In fact, I agree more than one-hundred percent. Technical Writing can be an arduous and frustrating path if you have no patience. I have written technical documents that have endured as many as twenty major edits/reviews. One Component Maintenance Manual (CMM) at EAM Worldwide took me six months to complete. Sometimes, I would work an entire day to create a one-page parts table supporting a complex technical diagram.

Patience is the supreme virtue that will enable and empower you to continue ahead successfully in this field.

Tip 245: Professionalism via Perseverance

Perseverance is the result of *applying patience* and combining it with insight. When writing a technical document, you should map out its milestones (major sections). Determine your information sources, review sources, and apply tools and resources as you step into the documentation project. In this way, you can *persevere successfully* and complete your technical document.

Tip 246: Professionalism via Sincerity

This writing tip is similar to what I stated about generosity. But, you can be generous without really caring. You can be generous but with a lacking attitude. The only way to turn generosity into authentic and positive generosity is through *sincerity*. My spiritual philosophy of "universal compassion" empowers me to write with purpose and keep my sincerity strong and at the forefront. *Sincerity* will enable you to endure SMEs who are having a bad day, managers who are having a bad day, misinformation, delays, and more. *Sincerity* comes from within. *Sincerity* will help you to cross the finish line and complete your almost impossible-to-write document.

Tip 247: Professionalism via Tenacity

The attribute of tenacity is great as long as you remain sincere. *Tenacity* is the most intense type of perseverance. However, you need to remain true to your purpose and your document for tenacity to work. You do not want to sabotage your efforts while working with others, who are providing technical inputs. *Tenacity* in a favorable sense will offer you inner strength and you will be able to write about topics that are complex and bring clear and positive meaning to the reader. *First, endeavor to understand. Second, endeavor as a writer.*

Tip 248: Professionalism via Zeal

Zeal is a word that is often seen in a negative light because it is frequently used to describe what some call "extreme religionists." But zeal is actually a good quality, if the focus is sincere.

Zeal means complete commitment to your work. Zeal means total enthusiasm.

When you write a complex technical document and are completely dedicated to this deliverable, you are going to take on each challenge with *zeal and determination* until your final deliverable is impeccable. Awesome!

Tip 249: Process Documentation

In my professional opinion, there are two major domains into which all types of documentation fit: *process* (procedural) and *summary.*

Note: This tip covers *process documentation* and the following tip covers *summary documentation.*

Process documentation is where you are instructing the user, step by step.

You could be presenting sequential steps in a work instruction where, for example, glue must be applied to a newly heat-sealed fabric so that, when inflated, it is strong and resistant. Perhaps you are helping a systems administrator to setup a new account on a web portal.

Both of these scenarios require step-by-step instructions.

Tip 250: Product Summary Documentation

The second major type of documentation is what I like to call [product] *summary* documentation.

Perhaps I have been tasked by my boss to write an external label for a new paint the company is creating and that will go onto the can. Here, I need to write out, clearly and correctly, each individual ingredient of the paint onto the visible label so that the customer can read it.

Perhaps my boss at a software company has tasked me with writing a short guide about Microsoft Planner. Before getting into tasks, I need to show the User Interface (UI) of Planner and here is where I compose a *summary* about what users will see when working with Planner and its diverse menus and sub-menus.

Tip 251: Project Management

Project Management is the practice of planning out an entire project from start to finish, touching on all details, milestones, resources, and strategies.

Project Managers (PMs) must know how to obtain, assign, and engage resources, both human and non-human.

PMs also need to compose documents that detail all the diverse aspects of their projects like strategies, decisions, tasks, metrics, and more. I have served many PMs in my career.

Project Managers empower diverse IT and business professionals, including Technical Writers, to work together so that a project is started and completed in good standing.

Tip 252: Punctuation Reminder

The English language features the following fourteen (14) punctuation marks:

Punctuation Mark	Symbol
Apostrophe	'
Brackets	[ ]
Colon	:
Comma	,
Dash	—
Exclamation Mark	!
Hyphen	-
Parentheses	( )
Period	.
Question Mark	?
Quotation Marks	" "
Semicolon	;
Slash (forward)	/
Slash (backward)	\

Tip 253: Punctuation (Apostrophe)

The technician disconnected the <u>server's</u> cable.

The DBA reviewed the <u>database's</u> performance metrics.

The programmer reviewed the <u>portal's</u> latest updates.

Tip 254: Punctuation (Brackets)

The [IBM] AS/400 is a midsize mainframe computer.

The [Apple] iPhone is the highest-selling mobile phone.

The [Miami] Herald is a popular South Florida paper.

Tip 255: Punctuation (Colon)

Open the following items: doors, windows, and vents.

Inform these three executives: CTO, CEO, and CFO.

Trust these computer brands: Dell, Acer, and Lenovo.

Tip 256: Punctuation (Comma)

After you finish the white paper, test the web portal.

When you meet with the SME, ask for his viewpoint.

Ask the supplies clerk for paper, pens, and pencils.

Tip 257: Punctuation (Simple Dash)

The 2017-2018 period saw many hurricanes.

Read pages 1001-1050 to learn what he said.

The company makes 1M-2M dollars each year.

Tip 258: Punctuation (Exclamation Mark)

Stop! That car is running the red light.

Go! You are next in line to greet Bill Gates.

Yes! The code passed review and goes live tonight.

Tip 259: Punctuation (Hyphen)

The <u>mobile-app</u> will be available for download soon.

The <u>process-analysis</u> showed a bottleneck here.

That <u>server-farm</u> is used by Google Cloud.

Tip 260: Punctuation (Parentheses)

Microsoft Office <u>(Word, Excel, etc.)</u> is great software.

Google Docs <u>(Docs, Sheets, Slides)</u> is cloud-based.

Both Toyota cars <u>(Yaris and Corolla)</u> are great rides.

Tip 261: Punctuation (Period)

C# is an object-oriented programming <u>language.</u>

New features must pass System Architecture <u>review.</u>

HTML is a markup language for <u>web pages.</u>

Tip 262: Punctuation (Question Mark)

Why did you discontinue that <u>vendor?</u>

Who is in charge of this <u>department?</u>

When will the software updates be <u>deployed?</u>

Tip 263: Punctuation (Quotation Marks)

Nike Corporation encourages us all to <u>"just do it"</u>.

<u>"Neo"</u> is a famous character in <u>"The Matrix"</u>.

Tip 264: Punctuation (Semicolon)

I will start the user guide tomorrow; today I test.

You favor proprietary apps; I favor open-source apps.

The CEO speaks English; the CFO speaks French.

Tip 265: Punctuation (Forward Slash)

The Human Resources Director should prepare his/her speech by noon for tomorrow's meeting.

The new platform gets rolled out, officially, on 1/1/20.

The Republican/Democratic initiative is receiving support from both the House of Representatives and the Senate.

Tip 266: Punctuation (Backward Slash)

The path to the file is C:\Docs\Books\2020.

Navigate as follows: Z:\SharePoint\GoogleChat. Here, you will see the application file you can download to your workstation. After downloading the app, install it. Then, you can use it to chat with other people on your team.

Tip 267: Purpose as a Reminder

Whether you are a Tech Writer or are someone writing a technical document, at some point in the journey you will feel the pain and discomfort of getting stuck.

You will need information. You will need input. You will need guidance. You will want to speak with SMEs. However, they will be busy and will not have time for you. *So, how do you proceed?* I say this: take this matter up with your manager or supervisor.

Stay dedicated and write down the original purpose and reason for this document. This way, your discouragement will pass. You will move on, eventually, and obtain your needed input. You will complete your tech doc!

Tip 268: Quality

In the business world, quality is a major concern. Without perfect or near perfect quality, products and services lose patronage. So, this is why companies have a "Quality" department of some kind. In software, you will see QA departments. In manufacturing, you will see QC departments. Quality departments work hard to ensure (1) company compliance with industry and (2) that products/services/apps leaving the company meet customer demand. Technical Writers will often find themselves supporting Quality departments by writing, editing, and proofing key technical documents.

Tip 269: Quality Assurance

QA (Quality Assurance) is a tech term you will hear frequently in the world of software development. QA engineers know how to test code and create test scripts. Technical Writers often support QA personnel because writing out test scripts can be challenging.

Tip 270: Quality Control

QC (Quality Control) is a tech term you will hear frequently in the world of manufacturing. QC engineers develop protocols and standards by which business production must abide. When auditors from a governing body like the FAA arrive, and sometimes without fair warning, it is the QC engineers who must present documentation indicating that the business is abiding by regulatory standards.

Tip 271: Readability

Readability is a term used to describe one's ability to reasonably comprehend presented information in a written document. Readability applies to technical documents as well as non-technical documents. Technical Writers should always aspire to attain one-hundred percent readability. Technical Writers should listen to feedback from SMEs, management, and readers so that all tech content is covered/addressed clearly.

Tip 272: Readers Perceive Intent

When I was younger (in my twenties) and learning about computer software and great applications like Microsoft Visual Basic and Microsoft FoxPro, I read technical books about these apps. I felt the positive intention of the authors of those books. Today, this is what I do daily. I make sure that my *writing intention is always positive.* Sometimes I state my intent in the document's introduction. Readers perceive your intent through the quality of your presentation. Take your time. Be clear and precise. When the reader completes your document and can then apply this new information well, you have succeeded as a Technical Writer.

Tip 273: Record of Revision

Including a *Record of Revision* page with your technical document is a practice that depends upon (1) your industry and (2) document protocols that your organization has in place. In the fields of software engineering and aviation, *Record of Revision* pages are recommended for guides of all kinds. Consult with your manager or supervisor regarding *Record of Revision* page protocols at your place of business.

Tip 274: Requirements

In the world of Software Engineering, requirements for software systems are akin to blue prints created by architects for houses that are to be built. Carpenters cannot get to work without such blue prints detailing the structure. Developers, QA engineers, and Technical Writers also will find it hard to work on a system without formal requirements. I have had several work experiences in my career where there have been no system requirements. Not easy!

If you have requirements, then you can positively benchmark actual functionality against the requirements document. *Technical Writers can write better user guides and training manuals with requirements documentation.* Do what you can to ensure that there are requirements for the system that you are documenting. If need be, offer to help your BA (Business Analyst) to write system requirements before you write the app's user guide.

Tip 275: Results, Not Always Visible

I have had managers ask me "Keith, how many pages of the user guide did you complete today?" For a software application that I understand fully and that requires no testing, I can produce as much as twenty solid pages of documentation a day. For other systems that are in development and that are full of issues, I might be lucky to produce four or five pages. I have to create meaningful diagrams or screen shots too, above and beyond text/content. So, on these days of minimal production, management sees no results, and the hard work I put into research is not visible or measurable.

In Technical Writing, you have to be fully committed to your work. It is hard for management to remain calm when you are only producing four or five pages of a user guide per day. *But little do they know that you are working like hell to investigate and verify information so that what you include on those four or five pages is both accurate and useful.* Persevere. The light at the end of the tunnel or the journey is a great user guide that will be used by many staff or external users.

Your manager will get great feedback and then say to everyone, "Thank you. The Tech Writer and I worked on this user guide together."

Tip 276: Reusable Documentation

If you are documenting a system that is constantly being revised, that is good news for you. Most of its core functionality will remain intact. Perhaps only a small percentage of its functionality will change or be enhanced. So, you can take previous write-ups of the system and reuse such diagrams and text in your new manual.

You might have to tweak the grammar and screen shots, but you can reuse a great majority of your documentation for the new software version.

Remember, divide and conquer. Write your documentation in small sections. This way, you can move topics or use cases easier with minimal rework. That is what I have done on several software documentation projects.

Tip 277: Rewriting

I once heard that writing success can be attributed to patience and perseverance in the rewriting process. I have witnessed this personally. It is true. Your documentation improves when you include feedback from management and subject matter experts (SMEs). It is hard for some workers and some managers to cope with numerous rewrites of a document. Still, if the company really wants great documentation, it needs to be revised constantly to reflect the system's current functionality.

Tip 278: Right Intention

My favorite book of all times is called "The Power of Intention," written by the recently departed Dr. Wayne Dyer. His books are all about authentic living and honest sentiment. This book is no exception. *Right Intention is what enables one to truly succeed.* There are numerous obstacles on the documentation roadway. The way you overcome these obstacles is through sincere dedication and intention. You have to love what you do.

I truly love what I do. That is why I succeed in writing high-quality user guides, gig after gig, year after year. *Right Intention* helps you to bring reason and sanity to the documentation project. *Right Intention* is your true power and positive force supporting your writing efforts.

Tip 279: Ronald Reagan

The 40th President of the United States was Ronald Reagan. He had a famous saying: "trust but verify."

I agree so much with this motto that I use it in my path of writing technical documentation. Let me give you an example. A BA gives me both business and system requirements documentation for a new platform that is being developed. I write the user guide based on the UI mock-ups pages provided and functionalities explained in this document. I finish the guide. Then, the system makes it to the test platform. I test it. I can then <u>verify</u> all the functionality presented/covered in the user guide. You should do the same. Trust your company and its information. Test when and while you can. Try to figure out the most important and applicable use cases (e.g. what the user will most likely be doing). If all of your testing confirms the requirements document info and the statements in your user guide, you are all set. If not, and you find something wrong, you are a hero because you have saved the company the embarrassment of deploying software that does not work correctly.

Tip 280: Screen Captures

The term "screen capture" refers to "screen shot."

It is important in not only user guides but also system requirements documentation and quality testing guides. Screen shots/captures should be included to accompany and enhance the meaning of the document's essential text. Readers develop more confidence in a document with screen captures than a document that is purely text.

Tip 281: SEO

The acronym SEO stands for *Search Engine Optimization,* which is the process and strategy of assigning metadata (keywords) to your document or webpage so that, when people search at sites Google or Bing, they will find your webpage/website easily. The best Content Management Systems (CMSs) have great built-in tools and resources for SEO. The developers of these systems do their best to see what search engines like Google and Bing demand in order to list a webpage favorably in search results. All Technical Writers should understand the basics of SEO.

Tip 282: Sharing Ideas

When I worked at EAM in Miami, Florida, I was part of a R&D project developing an escape slide for an aircraft client. There was no official Component Maintenance Manual (CMM) for this product, so I had to start from scratch. I held numerous meetings with engineers on the project and we shared ideas on how to meet FAA requirements as the CMM was written.

Tip 283: Showing What Systems Do

Technical Writers are empowered to show readers what a product/system can do. Your document is the way!

Tip 284: Sincerity

Technical Writing is more than just explaining in a direct and impersonal way how something works or what the parts of an assembly are. Technical Writing is the practice of applying written guidance and empathy. Through *sincerity* in your word choices, the more the reader will sense, that you are mindful of his/her reading experience and that you are writing to promote his/her success regarding the topic/topics you are covering. *Sincere documentation* is the bridge, metaphorically-speaking, between the left bank of the river (the platform, the technology, etc.) and the right bank of the river (the reader, the one who needs to understand the technology). Every well-chosen word makes that bridge stronger.

Tip 285: Small Steps

When writing documentation, the big picture exists. But, it only exists *in relation to* the total sum of individual steps that must be taken.

Let us say the user needs to print out a monthly activity report. The user guide explains the following major steps and individual steps therein: (1) logging in, (2) locating the reports menu, (3) choosing the right menu option, and then (4) entering the report parameters in the visible pop-up window so that the information pulled from the database and presented in the report is accurate.

Map out the overall journey. But, complete the journey *slowly and surely* through *mindful and correct* individual steps in your technical document.

Small steps help readers greatly.

Tip 286: Snagit

I have worked with several screen-capture programs over the years, and my favorite one is _Snagit_ by TechSmith. _Snagit_ features an intuitive yet dynamic and user-friendly interface. You can capture not only individual screens or parts of screens, but also more than one screen as you scroll down. So, if you have to document a report, for example, you can scroll down and capture the length of the report in one screen grab. Then, you can paste that screen grab into your user guide. Also, the resolution of the Snagit screen grab is fantastic. You can read all details of the image clearly. Great software!

Tip 287: Software Development

Documentation is an integral part of each stage of the software development process, also called the _Software Development Life Cycle_ (SDLC). Regardless of whether your company uses _Waterfall_ or _Agile/Scrum_ approaches (which I discuss later on in this book), there is still a "pipeline" present in the sense that the output is functional and useful code for the system. Requirements must be written by Business Analysts and Project Managers. Technical writers can help here. Then, Architects must review these requirements to see if they are viable, system-wise. Technical writers can help here as well. Next, these requirements are passed along to developers so that code can be written to meet functional expectations.

Technical writers support developers here by using tools like Sandcastle and/or Doxygen to document the source code being written. When the code is complete, technical writers can help QA engineers to write test scripts ensuring that the code performs as it should according to presiding requirements document. When the software is tested and ready for deployment, the Tech Writer then composes a guide for the user-base. _Tech Writers are key support personnel at every stage in the SDLC._

Tip 288: Sometimes Life Offers One Chance

A great piece of advice I got from my dad (when I was young) was "Son, sometimes life will give you just one chance. Do not let that chance go." I agree, still, with that advice. My wife and I adopted our dog, Ben, in 2010 from a Broward County (Florida) animal shelter. At the time, we were new house owners and wanted to have a dog at the house. So, we went to the local shelter and met Ben. The woman there told us "He is a really nice dog. If you do not adopt him right now, the next people in line will probably adopt him". So, we decided right then and there to adopt Ben. He has been our dog now for eight years. He is a great addition to the home. We need to learn to make key decisions on the spot and the more mindful we are, then the better the decision we will make.

Tip 289: Sometimes Life Says No

There is an ancient Buddhist teaching that says something like this: "if your cup is full, you cannot pour fresh tea into it." You must first empty the cup so that you can then pour new tea into it. There is great wisdom and truth in this teaching. On several occasions of my IT career, out of the blue, my Technical Writer position was closed. There was a restructuring of personnel, for example, and some of the less critical roles like Technical Writer and Technical Trainer were removed from the roll call. It is easy to see a time like this as a loss.

But let us not forget the Buddhist teaching, as shown above. In a short amount of time, after these times, I started working at even greater companies with greater software development projects needing documentation and training. *Life is a continuum.* We like to "snapshot" things and look at life sometimes through a single moment. But, life goes on and things change. So, if life says "no" right now, fear not! The universe is preparing something great for you. *Great projects are manifesting for you.*

Tip 290: Spacing

In Microsoft Word, for example, you can verify and/or modify the line spacing in your document by placing the cursor in your document and then by right clicking on your mouse. A pop-up window displays and in this pop-up window you can see the current line spacing settings. You can modify these settings as you need.

Tip 291: Standard Operating Procedures

In both universes – technology and business – SOP means *Standard Operating Procedures.* This is a common theme among user guides. Let us say I am tasked with writing a SOP guide that will be used to train forklift drivers. A forklift is an industrial truck powered by electricity or gas that features a mechanical-lift on its front part, so that it can lift/move heavy loads. Factories, warehouses, and assembly plants usually have forklifts and employ forklift drivers. So, the *SOP* manual will have to cover use of the forklift, its most important components, and also forklift device operating rules mandated by the company.

Most likely, the SOP manual will have to be approved by the industry's auditing authority and, in this way, the forklift driver is operating in compliance with both company/industry rules and regulations.

Tip 292: STC

The Society for Technical Communications (STC) is the largest and most longstanding professional association that is dedicated to the advancement of Technical Communications, especially Technical Writing. I value what the STC has to say. At the same time, in the business world, important technical documents have to pass audits and reviews of governing bodies like the FAA, ISO, and/or FINRA, just to name a few. So, at the end of the day, the mandates of these auditors are slightly more important than those of the STC. If your technical documents do not have to pass such audits, however, then you can use the STC as your formal point of reference. I have gotten to know several people from the STC over the past decade and they are all outstanding individuals and Technical Writers.

KUDOS to the STC. Keep up the great work!

Tip 293: Staying Motivated

Technical Writers need to stay motivated in order to stay productive. The best way to stay motivated is to periodically stay in touch with your reader base.

When I recently worked at 1GC, one of the company's "preferred customers" was a local and immediately accessible sales companies. Periodically, I would share early drafts of user guides that eventually this company would be using. I received positive feedback from senior managers. Then, I used these comments to improve and better my technical documents. You write to help others. Your documents will help them to accomplish key tasks. *Touch bases from time to time with your readers to stay motivated.*

Tip 294: Success Principles, Practices

I recommend you grab a copy of a recent book I wrote called "Success Principles and Practices" that is now live at Amazon.com. It features great teachings of the Buddha about *authentic success*. The success principles featured in this book are rock solid and they have helped me get to where I am today in my life and career. Success principles and *virtuous practices* like patience, perseverance, and sincerity are all true empowerments toward Technical Writing mastery.

Tip 295: Table of Contents

The general rule that I follow and that has been mandated to me by many companies over the years is this: documents that are nine pages or less do not warrant a Table of Contents (TOC). However, TOCs are necessary for documents that are ten pages or more.

Creating a TOC is simple. Make sure that you configure your headers – Header 1, Header 2, etc. as needed in terms of font size, bold, italic, etc. and then after you have written your document, and selected the items that are to be headers, return to the page where you will feature the TOC. In Microsoft Word, go to References > Table of Contents and select the TOC option best for your document. Word will create the TOC based upon the headers you have designated throughout your document.

Tip 296: Teamwork vs. Solowork

The best way to approach Technical Writing is with an optimistic spirit. In some work environments, you will have an infrastructure – co-workers, management, SMEs, and others who are technically-proficient. Here, you can implement and apply team work methods, strategies and milestones.

In other work environments, you will have no infrastructure. It will be just you, alone. Still, you can create a document that maps out your writing journey and work hard to attain those milestones.

Teamwork is great because you have feedback from others. Sometimes that feedback is insightful, other times not. Still, a different set of eyes on your document is helpful, for the most part. *Solowork,* on the other hand, requires more self-discipline. You need to be able to switch intellectual modes from writing to editing, for example. Make the most of the situation you have at hand. Then, you will write your tech docs with great success.

Tip 297: Tech vs. Finance

In the same way that a marriage today is between two consenting adults, business also is a marriage of Technology and Finance. Generally, Finance is the key stakeholder because it injects and manages capital, which allows for a business infrastructure to exist and for the pursuit of profit to take place. Technology, on the other hand, is the second "person" in this business marriage model. Technology is in charge of the business "technical infrastructure" like company phones, servers, email, software, training, power management, disaster recovery strategies, documentation, information, and more.

At the end of the day, all major business decisions are the result of both Business & Technology inputs.

Tip 298: Technical Writing, Worth It?

Technical Writing can be an arduous professional path.

Sometimes, there is little recognition. You write technical document after technical document. There is no mention of who is now reading it. There is no thank you from supervisors, management or business owners.

Are you working in vain? Absolutely not! You are helping your readers. Your understanding of a certain system, software or process is helping them to attain new levels of productivity and technical competence.

Technical Writing, most certainly, is worth the effort!

Tip 299: Ten Google Search Tips

For the following examples, enter info (into the www.google.com Search box) and press Enter:

- AND: Enter Mustang cars 1969 AND 1970
  Note: Results show Ford Mustang cars for 69,70.
- Asterisk: Enter * makes the world go around
  Note: Results show first word of this saying.
- Define: Enter define algorithm
  Note: Results show words defining 'algorithm'.
- Double Qs: Enter President Teddy Roosevelt
  Note: Results show info about President T.R.
- File Type: Enter filename.pdf
  Note: Results show the URL for this PDF.
- Itinerary: Enter Itinerary Chicago Boston
  Note: Results show flights, Chicago/Boston.
- Phone Numbers: Enter 9547777777
  Note: Results show Broward Cab info.
- Site: Enter site www.techwriterguy.com
- Temperature: Enter temperature, Miami
- Package Tracking: Enter a package tracking #
  Note: Results show shipping & transport info.

Tip 300: Tenses for Verbs

Overall, in the English language, there are <u>six tenses</u> that apply to verbs. They are as follows: *past, past perfect, present, present perfect, future,* and *future perfect.* Each of these tenses has a simple and progressive form. Here is one verb that I frequently use in software guides:

<u>CLICK</u>: here are the verb's different forms by tense:

Tense	Simple Form	Progressive Form
Past	I <u>clicked</u> on the Run button.	I <u>was clicking</u> on the Run button.
Past Perfect	I <u>had begun clicking</u> on the Run button.	I <u>had been clicking</u> on the Run button.
Present	I <u>click</u> on Run button.	I <u>am clicking</u> on the Run button.
Present Perfect	I <u>have clicked</u> on the Run button.	I <u>have been clicking</u> on the Run button.
Future	I <u>will click</u> on the Run button.	I <u>will be clicking</u> on the Run button.
Future Perfect	I <u>will have clicked</u> on the Run button.	I <u>will have been clicking</u> on the Run button.

Technical Writers and Technical Communicators should do their best to use <u>Present Tense</u> in its <u>Simple Form</u>.

<u>For example</u>: *Click* on the *Run* button. The IDE *compiles* the program and then *displays* the results at the bottom of the visible window.

Tip 301: Test-Driven Development

Technical Writers who serve the Software Development industry should know that *Test Driven Development*, also called TDD, is a popular coding and testing approach, especially in *Agile/Scrum*, where one QA resource is paired with one Dev (development) resource. Technical Writers sometimes are asked to write summaries for the QA engineer about what he/she has found in testing the developer's code. Here is a great article about TDD:

http://agiledata.org/essays/tdd.html

Tip 302: The Golden Rules

The *Material Golden Rule* says that whoever has the gold makes the rules. In today's business world, it is investors who frequently drive businesses and their direction. They want their investments to produce a positive return.

The *Spiritual Golden Rule* says that one should be kind and treat others in a way that one also wants to be treated, which generally is based on mutual respect.

Both rules are important:

Respect upholds the first golden rule.

Respect upholds the second golden rule.

Tip 303: Think Globally, Act Locally

GLOCAL, which means to think in a global sense but act in a local sense, is a term from the 2000s when the Internet was growing and expanding and businesses were also getting onto the web in some way. It is still a positive and applicable idea today. Technical Writers should understand this idea because many systems have a larger and common purpose; but documentation needs to show

how individual users can benefit in different ways according to their local circumstances and unique uses of the software.

Tip 304: Time is Valuable

We should all respect one another's time. Most people today are hard-working middle-class citizens and so they do not have a great deal of free time. Many of them work six or seven days a week earning needed money and they are also busy taking care of their families and communities. Respecting others' time is a way to promote positive human relations during challenging times.

Tip 305: True Teamwork

Teamwork is a great blessing. To be part of a team where everyone is fully dedicated is an amazing and empowering experience. Teamwork increases the likelihood of success. Teamwork allows for team members to find answers to questions and for progress to be made but also in light of other activities of the team. *Technical Writers should embrace teamwork.*

Tip 306: Udemy

If you wish to study writing for either free or almost-free, look no farther than www.udemy.com. This website has thousands of free and almost-free business, technology and education classes online.

Tip 307: Udacity

If you wish to complete a high-tech online class and then be truly marketable in the eyes of companies like Google, Facebook, Amazon, and Apple, consider www.udacity.com, an amazing online tech-education site started/created by former Google executive Sebastian Thrun.

Tip 308: UI

UI stands for User Interface. UI addresses web pages, pop-up windows, buttons, text boxes, radio buttons, and essentially how the user interacts with the program or webpage. Tech Writers should familiarize themselves with all the different tools that developers use while creating a User Interface.

Tip 309: UX

UX stands for User Experience. UX is similar to UI but is more concerned with process/work flows, tasks, interests, and needs of the user (the big picture). UI, in my professional opinion, is a sub-domain of UX. Still, both are critical to creating great software, whether it is a desktop app, web app, or mobile app. Technical Writers should familiarize themselves with all the different kinds of tools and resources that developers use while creating a UX.

Tip 310: Understanding Code

Today, the most important markup languages for Technical Writers to understand are HTML (Hyper Text Markup Language) and XML (eXtensible Markup Language) and this is because a great deal of software documentation ends up on the web or an intranet.

Tip 311: Understanding Gerunds

In the English language, a "gerund" is a unique form of a verb that serves as a noun and that also must end in the three letters "ing".

Examples: working, typing, studying. The verbs here are work, type, and study. However, to transform these verbs into nouns, you must apply the three letters "ing" at the end and then use this new form as a noun.

Here are some sample sentences:

<u>Working</u> long hours is common to those in IT.

<u>Studying</u> is the best way to prepare for final exams.

<u>Typing</u> is a task well-known to Technical Writers.

<u>Tip 312: Understanding Iteration</u>

Iteration, apart from specific context, means "repetition". Applied to software development, "iteration" means "new version". In *Agile/Scrum,* where there is a regular two or three-week "sprint", or mini-cycle to produce working code, iteration is the sum result of the "sprint".

At the end of the "sprint", shippable/deployable code is released into production for the benefit of the overall system. There could be code that affects the user interface and there could also be code that simply enhances the efficiency of the system but there is no visible difference in the UI. Technical Writers should understand how iteration will affect their system documentation.

<u>Tip 313: Understanding Magic</u>

"Magic is believing in yourself. If you can believe in yourself, then you can make anything happen." This saying is from Johann Wolfgang von Goethe. I agree!

<u>Tip 314: Understanding nTier</u>

Technical Writers should understand at least the bare bones of "nTier". This expression means "multi-level" in terms of system functionality.

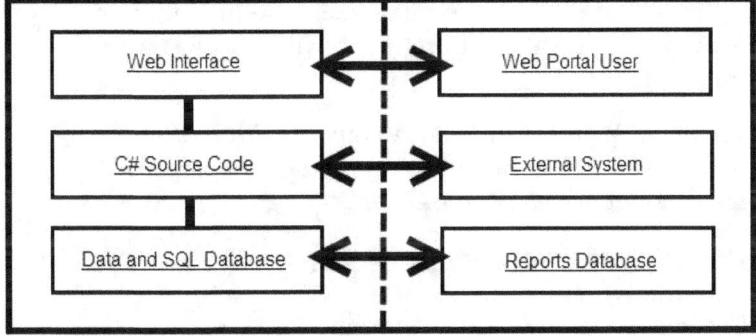

Typical nTier computer systems have the following tiers that all have unique features, characteristics, and purposes: (1) data, (2) business logic/rules and main system/software code, and (3) user interface (UI).

For software companies that do not have extensive resources, they might end up asking the Technical Writer on staff to document all levels/aspects of an nTier system.

Tip 315: Understanding Scalability

To scale a business means to "grow" or "expand" a business. Let us take a software company that creates a printer driver. The driver sells well and the company decides to expand into other areas to create printer drivers for those markets. The company needs to invest in more human resources, workstations, telephones, computers, servers, etc. The ability of the company to increase its monetary investment and yet successfully produce revenue is the secret to "scaling" a business.

Tip 316: Understanding Transparency

Transparency means "invisibility". In the business sense, it means "invisibility" regarding the creation of standardized code, processes, procedures, logos, document templates and more. Surely, great corporations like Microsoft & Google work hard to enforce deliverable transparency. All logos, documents, communications, styles of code, and more must seem like it is just one person behind the curtain, if you will. That is true transparency.

Technical Writers must learn to work with written communication standards to uphold an employer's product/process transparency.

Tip 317: Understanding Use Cases

Use cases are end-user scenarios.

All of the tasks that the typical end user faces can be turned into documented *use cases*. Technical Writers can, at least in requirements and testing documentation, itemize *use cases* and in this way complete a writeup that addresses all system functionality, feature by feature.

Here are a few *sample use cases*: (1) logging in, (2) changing one's password, (3) changing options within one's settings page, (4) generating a report, (5) entering data and getting an email confirmation, (6) uploading information to a portal, and (7) downloading a PDF.

Tip 318: Unity is Strength

An ancient Swahili saying declares that "unity is strength". I agree one-hundred percent.

Most Technical Writers today end up on a team of some type. The team spirit is what allows for optimal participation, utilization of time and resources, and greatness in documentation. *But this should not be done in spite of one another.*

Rather, projects need to be managed so that tasks are *concurrently accomplished* and *unity works* across the team in positive and supportive ways.

Tip 319: User Guides

"User Guide" is the most common name attributed to the technical document that shows end users key aspects of a software app, system, or platform, from basic user tasks to intricate/complex usage.

My recent book "Windows 10 for Seniors and Beginners" is a user guide showing individuals new to Microsoft Windows 10 what the most important User Interface (UI) elements are as well as how to accomplish important tasks like managing Settings.

Technical Writers should work closely with Architects (system side) and Business Analysts (logic side) to ensure that all important system topics are addressed.

It is one thing to simply show your reader how to do something, in the user guide. It is better to show that functionality in light of why that functionality is necessary.

Tip 320: Using Diagrams

Diagrams are a powerful way to shed light on important text in your technical document.

At a recent work engagement, I created an *Onboarding Guide* for new staff members. I created a diagram similar to this one (but far more detailed) showing the overall business model of the company (and many people thanked me for writing this informative guide):

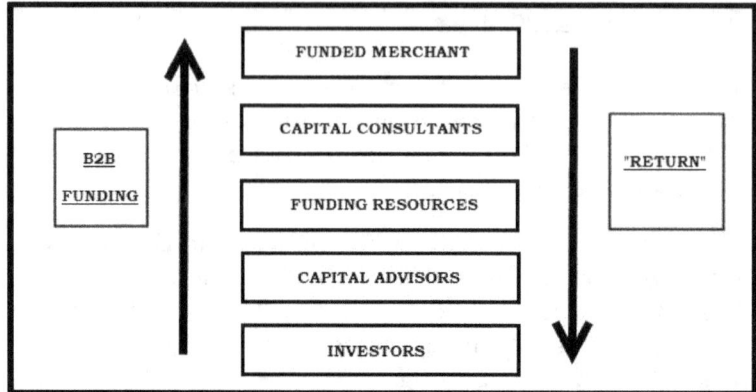

High-quality diagrams will help you, as a Technical Writer, to increase meaning and inject new comprehensive value and readability into your document.

Coordinate diagram creation and inclusion with management so that such diagrams meet executive expectations.

If your company (or you) can afford the software, I highly recommend *Microsoft Visio.*

If you or your company are on a shoestring budget, however, then use a free diagramming software like *yEd.*

Here is its webpage:

https://www.yworks.com/products/yed

Tip 321: Verbs, Intransitive

There are two classes of verbs in the English language: transitive (which I cover in the next tip) and intransitive. Verbs that are intransitive <u>have no object</u>. However, they can be followed by a prepositional phrase. Here are a few *intransitive verbs* used in tech-related sentences:

The laptop battery <u>failed</u> because it was not recharged.

The server <u>died</u> because it suffered a hacking attack.

The CTO <u>sat</u> in his chair before attending the meeting.

Tip 322: Verbs, Transitive

Transitive verbs in the English language <u>have an object</u>.

Examples (I have highlighted both the verb & its object):

The candidate <u>will receive a job offer</u> yesterday.

The IT support analyst <u>is configuring new laptops</u>.

The programmer is <u>writing new code</u> for the system.

The DBA <u>backs up the company's data each Friday.</u>

The technical trainer <u>delivered his presentation</u>.

The CEO decided <u>to implement a new phone system.</u>

The CTO now <u>authorizes employees</u> to wear jeans.

The Tech Writer <u>has completed the user guide</u>.

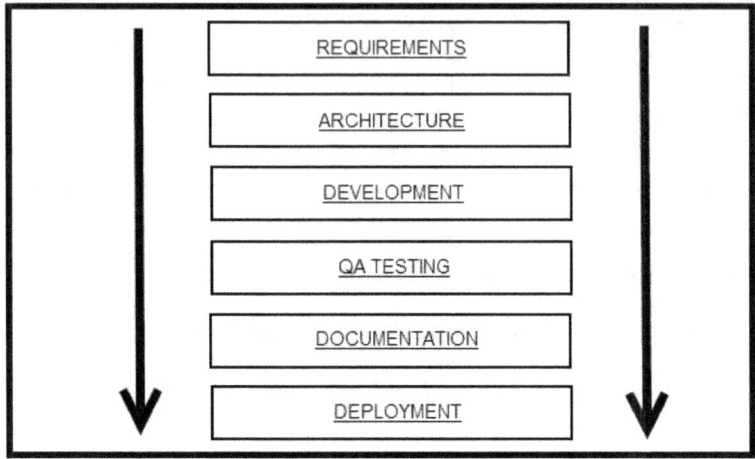

Tip 323: Waterfall

Waterfall is the name attributed to the software development process used before *Agile/Scrum* came along.

Waterfall is an approach to software development still in use by many organizations and especially those of a smaller size. Waterfall does not work with iterations in the same way that Agile/Scrum does. Waterfall is a linear pipeline software development path. Here are the major steps and milestones of the step-by-step Waterfall software development path:

(1) Requirements Documentation
(2) Architecture Review/Editing of Requirements
(3) Software Development Process (e.g. Coding Process)
(4) Quality Assurance testing of new code
(5) Documentation, especially User Guides
(6) Deployment

Tip 324: We Are What We Do

The ancient Greek philosopher Aristotle once said that "We are what we repeatedly do. Excellence, then, is not an act, but a habit." I agree. In order to become excellent at something – an art, a craft, a hobby, or a profession - you need to endeavor with great *diligence* and *perseverance*. This book, for example, is the result of twenty-years of dedicated work in the field of Technical Writing.

Do not try to do anything massive in just one day. Give yourself time and pace yourself. Life is a journey. Your career is a journey.

You will attain your goals and reach your destinations. You will be recognized for what you do.

Tip 325: Web Browsers

For all of your [researching] work on the web, I recommend *Google Chrome* as the browser of choice. It is safe and yet powerful and resourceful. However, if you are not a Google fan, then I recommend *Microsoft Edge*, which is a close second, in my professional opinion.

Here are their respective download pages:

https://www.google.com/chrome/

https://www.microsoft.com/en-us/windows/microsoft-edge

Tip 326: Whiteboards

Whiteboards are excellent tools for meetings, concept brainstorming, and idea development. Whiteboards are able to be marked up extensively and at the same time erased, fully or in part.

Whiteboards enable many different people to contribute to a drawing or plan or outline together, for example, and at the end of the whiteboard-centered meeting, the result is right there in plain sight for all to see.

Get a whiteboard for your business or classroom. Use this whiteboard extensively. You will see the quality of your team's work improve greatly.

Tip 327: Wireframe Software Tools

The purpose of "wireframe" software tools is to enable users to focus on an app's user interface without having to worry about writing code. As I mentioned before, a typical system today has three major layers: data, code, and user interface.

Wireframe tools work with the UI.

Two tools that are worth mentioning, here, are *Evolus Pencil* and *Just in Mind*.

Both are great wireframe apps.

Here are their respective web addresses:

https://pencil.evolus.vn/

https://www.justinmind.com/

Tip 328: Who vs. Whom

The word "who" refers to the subject of a sentence, while the word "whom" refers to the object of a verb or preposition.

Examples:

I am a person <u>who</u> values hard work.

Note: Who describes "person".

The PM is hiring individuals <u>who</u> know Microsoft Project.

Note: Who describes "individuals".

The person <u>to whom</u> you are grateful is the CEO.

Note: Because you can rewrite the sentence using him/her, the use of "whom" is necessary. For example, "you are grateful to him (e.g. the CEO)."

Tip 329: Who, What, Where, How, When, Why

Technical Writers have six key *question types* that can be used when researching information: (1) who, (2) what, (3) where, (4) how, (5) when, and (6) why.

Here are some examples.

<u>Who</u> wrote the code for the new application?

<u>What</u> kind of application was developed?

<u>Where</u> is the code located (so that I can review it)?

<u>How</u> did the developer convert requirements into code?

<u>When</u> did the developer complete the coding task?

<u>Why</u> was this developer assigned to this project?

Tip 330: Working with Acronyms

Acronyms are abbreviations where each letter in the acronym represents the first letter of a different word, for a short sequence of words.

Examples:

DOS stands for Disk Operating System.

BIOS stands for Basic Input Output System.

CTO stands for Chief Technology Officer.

In your technical document, always state the complete series of words along with its acronym in parenthesis at first mention. Then, for the rest of your document, you can use the acronym because you have already explained its meaning at the outset.

Example: "Our Chief Executive Officer (CEO) will be dedicating an entire wing of the building to the Project Management (PM) department."

Tip 331: Working with Antonyms

Antonyms are words that are the exact *opposite* of a given word or term. The antonym for "light" is "darkness". The antonym for "open" is "close". You can use antonyms to *positively contrast* a situation.

Example: "The IT Administrator must _open_ the laptop and set it up with Microsoft Windows 10, Office 2016, and Visual Studio 2016. He/She must then run Windows Defender. When that is complete and the laptop is configured, he/she must _close_ it and place it on the new employee's workstation".

Tip 332: Working with Accountants

The next twenty-or-so tips are important. They are not about writing or the English language. Rather, they are about *working with people*. This is a skill that is starting to wane in light of the emerging mobile phone generation. Strong people skills ensure productivity, in my book. First in line are accountants. They are factual and numbers-based. Do not try to sell an accountant on an idea. Rather, sit down with an accountant and map out your idea on paper and show how your approach makes logical and mathematical sense. The accountant will most likely review your diagram or documentation idea with great detail. You will get feedback in the form of diagram corrections, improvements, and more. If only your regular SMEs were so detailed! Now, you can get to work writing that new SOP document for the Accounting Department. Now, you can write a solid user guide about the company's new accounting software, for example.

Tip 333: Working with Attorneys

Attorneys, like Technical Writers, are usually gifted writers, thinkers, and have a strong command of the English language. So, when working with Attorneys, work closely to ensure that your document is strong grammatically as well as presentation-wise, using the correct language *mandated by law* that pertains to the document you are writing. You could be proofing a lease. You could be co-authoring a sales contract. Remember that you do not possess the legal knowledge that the attorney possesses. Still, you can ensure that the document's grammar is strong and that the presentation is consistent and readable. If you do not understand specific words, ask about them and the attorney will verify their accuracy and if they are needed in the document.

Tip 334: Working with BAs

In my Technical Writing career, I have noticed that the top task of Business Analysts (BAs) is *requirements gathering and consolidation.* BAs work hard to create software system requirements [documents] which are then used by diverse developers, database analysts, technical writers, trainers, and others. Requirements for software are akin to blue prints or plans for a house. They are important. BAs are under tremendous pressure most of the time, so be patient with them and if you know of a resource that will help them to complete their current document, they will be grateful. BAs are process-minded so as you work with them, your diagramming skills will increase greatly.

Tip 335: Working with Collectors

Collectors have the daunting task of asking customers and clients to pay up. Usually, when a collector calls a client, it is regarding outstanding payments and/or debt on an account. So, collectors need to integrate strong financial skills with strong communication skills in order to accomplish their collection tasks. Technical Writers can help these professionals by creating clear-cut process flows showing different psychological and/or financial strategies to apply at different stages in the overall collections process.

Like BAs, Debt Collectors are process-minded. Keep that in mind when working with these brave and courageous financial professionals.

Tip 336: Working with Developers

I have spent the majority of my professional career working with computer programmers and software developers. In short, they are brilliant individuals. They have a powerful skill of being able to transform system requirements and ideas into functional code. If that is not a gift, then I do not know what is. Technical Writers cannot master all of the languages that developers use. However, Tech Writers need to be able to document source code when working with developers.

When working with Microsoft Visual Studio, you can use *Sandcastle* to document code written in this IDE. When working with Eclipse, which is used for Java development, you can use *AgileJ StructureViews*, which is an add-on for Eclipse. If you are a believer in open-source, look no farther than *Doxygen* written by Dimitri van Heesch. *Doxygen* is amazing, as it works with C++, C#, Java and many other major programming languages. I used Doxygen when I worked at Qpay and worked on the Architecture team. New programmers consulted with the *Doxygen*-generated C# class hierarchy diagrams and code comments (that I compiled) to get up to speed quickly.

Tip 337: Working with Dogmatic Thinkers

One of my favorite humanitarians is the 14th Dalai Lama of Tibet [now the Tibetan Government in Exile], Tenzin Gyatso. I have seen him speak twice here in South Florida. Here is what he said, on both occasions, about dogma (I am pulling this quote from memory): "Dogma means rules. We need to follow rules to live in peace with one another. Still, there are times when the rules get in the way of peace, and so we must break these rules very carefully."

Let me give you a practical example. There is a hacking attack on a server. There are two IT Administrators on staff – a Senior Admin and a Junior Admin. Only the Senior Admin has the authority to log into the Server

Management Suite (SMS) and perform Admin tasks. But, at the time of the attack, only the Junior Admin is on-site. There is no time for the Junior Admin to call the Senior Admin. It is now or never to block the attack. So, without permission, the Junior Admin logs in to the SMS and knows what to do because he has watched the Senior Admin defend against attacks previously. The Junior Admin saves the Server and the entire corporate database. Dogmatically, he is wrong. He broke corporate rules. But, on the other side, he saved the company millions of dollars in breached data.

This is an example where the CEO and CTO should thank the Junior Admin for taking the initiative as opposed to firing him for breach of protocol, in my professional opinion. If the Junior Admin had ample time to call or text the Senior Admin, then that is a completely different scenario. But, here, the Junior Admin had zero time to react. It was "react or suffer a server breach". This courageous IT Junior Admin ended up saving the IT team many hours of remedial work.

Conclusion: Dogma is helpful 99.9 percent of the time. But for 0.1 percent of the time, we need to apply *common sense* to make the best possible decision.

Tip 338: Working with Dreamers

A "dreamer" is a visionary. The world needs visionaries. Just about all of our incredible modern inventions started off as mere dreams. The *automobile* was once a dream. The *airplane* was once a dream. The personal computer was once a dream. *Air conditioning* was once a dream.

Today, through great engineering, we have many automobiles from which to choose. Today, we can choose among many commercial flights between Miami and New York City. I can go to Best Buy and select the right laptop personal computer for me.

I can also choose among hundreds of air conditioning vendors and systems and implement the one that is just right for my home.

Dreamers matters. At the same time, they need to learn to work with best business and engineering practices, regardless of industry.

In this way, the idea of the "dreamer" can become a real product or service that can become both a prosperous business as well as serve/support its customer base.

Tip 339: Working with Engineers

For two years, I worked at EAM in Miami, Florida and was a Technical Writer for an important R&D project.

I wrote the Component Maintenance Manuals (CMMs) for this R&D project as well as supported the team's mechanical engineers in all of their tasks. While working with engineers, like accountants, details are everything. An idea or process or "conclusion" is accepted only as long as relevant facts and information are able to be validated.

So, working with engineers means taking a step back from human politics and focusing on actual parts, numbers, angles, testing results, assemblies, and things like these, both individually and collectively.

In the world of engineering, you work with OEMs (Original Equipment Manufacturers) extensively. It is these companies that have the latest information about their products and product assemblies that engineers use/utilize. *This* is the information you need to bring to the table and discuss with engineers who work with these products. Engineers will tell you how to understand these diverse products and how to incorporate such details into your user manuals.

Tip 340: Working with Executives

In the same way the Coast Guard has icebreaking ships in frozen oceans through which ships need to pass, businesses/organizations have executives who are there, literally, to forge the company's business path. Executives are not usually involved with day to day operations. Rather, they are involved with venturing into the unknown to gain business contacts, contracts, deals, and more.

On a few occasions, I have helped executives to write key documents for meetings, presentations, business expos, and town halls. The two most important words that come to mind, here, are *brevity* and *consistency*. Executives can always point interested clients or customers to specialists within the business for more details.

Executive documents and presentations need to be incredibly *specific* because their time is limited and they know this. Tech Writers can help executives with critical documentation that can help the business grow.

Tip 341: Working with Graphics Specialists

Over the course of my Technical Writing career, I have worked with numerous graphics specialists, CAD specialists, SolidWorks specialists, front-end developers, and individuals with a keen ability to turn chicken-scratch lunchtime-sketches into sophisticated and complete diagrams and drawings.

Thank you all for this great inspiration.

To work with this gifted class of professionals, the magic keyword is "time". It takes *time* to create. It takes *time* to modify an existing diagram. It takes considerable *time* to redo the view so that you see a part, for example, from above, from below, from the left, and from the right. Technical Writers might think "how long can it take to make just a single-page assembly diagram?".

It could take one day. It could take one week.

Make sure that you work side-by-side with your manager (who is waiting for your finished document) and graphics team so that it does not look you are sitting idle at your desk. Ask the graphics specialist(s) what you can do to expedite the finished work. Be patient. The diagram you need for your tech doc will eventually be completed.

Tip 342: Working with Hardware Specialists

Documenting hardware functionality is precarious. It is much more complex than documenting software. Hardware involves all sorts of complex parts and part interactions. You must work closely with experienced technicians (e.g. hardware specialists) as well with contacts at your Original Equipment Manufacturers (OEMs). In this way, you will be able to identify functionality correctly as well as cause and effect. In software, it is not too hard to verify specific functionality. In hardware, that is not always the case. You could see a light turn on, but that does not necessarily mean what you think it does. Again, work closely with hardware specialists and experts and OEMs to ensure that every single word in your documentation is spot-on accurate.

Hardware documentation has zero percent tolerance regarding error/non-accuracy. I highly recommend you work with at least two or three separate SMEs who are experienced hardware specialists to make sure your document is perfectly written.

Good luck!

Tip 343: Working with Management

The key to working well with management is what I call "deliverable alignment". Eventually, you report to one person only. If anyone approaches you asking for help or input into their technical document, that is ego-pleasing for sure. Still, your manager is waiting for important documents and you do not want to unnecessarily delay such deliverables. Tell the person needing help to take it up with your manager. If it is important, your manager will work it into your pipeline. If not, well, then your co-worker will have to find other eyes with more time.

Tip 344: Working with Negativity

Negative energy is toxic. It troubles me to see negative people. I empathize with afflicted individuals because they still do their best to perform at work. Negative people do not try. Keep your distance from negativity.

Tip 345: Working with Non-Technical Users

A major challenge I have faced in my career is having a highly-technical boss like a Systems Architect (SysArch), Project Manager (PM), or Senior Business Analyst (SBA) and he/she tasks me with writing a user guide for a non-technical audience. This audience could be internal. This audience could be external.

Many times, these individuals forget that the reader knows about one-percent of what they do.

So, I have to remind them "boss, you know the whole system. The reader works in customer service and needs me to spell out things that you think are not important. Please do not delete my observations, comments, and notes. My manual is needed training as well as needed documentation." Advocate for your non-tech readers.

Tip 346: Working with Organizational Rules

In the movie "Top Gun" featuring Tom Cruise, there was an incident where Pilot Pete "Maverick" Mitchell (played by Cruise) buzzes the Top Gun School Control Tower. In response, the School Director tells him "the rules of Top Gun exist for your safety and that of your crew. They are not flexible nor am I."

Rules established by businesses, similarly, are for the benefit of all who work there. I do my best to respect organizational rules knowing that in due time I will see why such rules exist. Organizational rules are not oppressive. Rather, they exist to help the company conduct daily business.

Technical Writers sometimes will find themselves helping HR to write documents about corporate rules/policies, like SOP guides/manuals.

Tip 347: Working with Positivity

Positive energy is awesome.

However, it is so powerful that it needs to be managed. Positive energy needs to be channeled. In this way, good vibes provide workers with energy and stamina necessary to produce high-quality deliverables and also work through challenging moments.

Technical Writers need to discover for themselves how to take their positive energy and put it all into a document. I have done this for many years now. First, I put on some inspiring music and make sure I have access to all documents and/or webpages I need that are inputs into my deliverable. Then, I get to work.

Tip 348: Working with Stakeholders

In *Agile/Scrum* software development, the *Product Owner* is essentially the *Business Stakeholder* while the *Scrum Master*, *Developers*, and *QA Engineers* are all technical experts who contribute to the business mission through their technical roles, skills, and aptitudes.

I obtained *Certified Scrum Product Owner* (CSPO) certification from *The Scrum Alliance* in 2013 through a course given by Jeff Patton, who is a nationally-recognized expert on Agile/Scrum software development. The class was called "Passionate Product Ownership". I have been able to take knowledge acquired from this class and successfully apply it to my Technical Writing career since that time. I encourage all Technical Writers to learn as much as possible about *Agile/Scrum* and its iterative path to software development.

Tip 349: Working with SMEs

SME is an acronym for *Subject Matter Expert*. I enjoy working with SMEs. In short, they are brilliant individuals and an entire of universe of knowledge resides within them. In order to access that information, you need to treat SMEs with great respect. Usually, they have important roles at companies; their time is precious. Sometimes they barely have time to read/respond to emails. So, you must work with your manager or supervisor in order to gain access, *legitimately*, to a SME.

Once you schedule a meeting with the SME, have your questions ready. Make sure you do as much legwork as possible beforehand. That way, your meeting time is productive. I always ask SMEs, too, for their input on how I can prepare before meeting with them. In this way, I have shown them respect and when they finally meet with me, they know that I am ready and able to work with the technical information they are presenting.

Tip 350: Working with Survivalists

Survivalists are not just people preparing for the end of the world. They are also individuals in business who are super-cautious and super-frugal to the almost-point of disrupting productivity and progress.

I have worked with a few such individuals during my career. Here is my takeaway: show a survivalist that you agree with being cautious. Explain to the survivalist that you are asking him/her to take just one step, only, out of his/her comfort zone. In other words, you value their cautious outlook and concern.

Survivalists are positive people. However, my guess is they have endured a traumatic event in their life and are just a bit defensive at work. Many times, they are brilliant and insightful and do not want to see their efforts go in vain. Technical Writers, in such a scenario, need to be *positive* in word (document) and spirit (speech). In this way, you can gain the trust of those with a survivalist outlook in business. These super-cautious individuals will then support your technical documents.

Tip 351: Working with Synonyms

Synonyms are words that carry the *same meaning*, for the most part, as another word.

Enterprise is a synonym for *business. Connection* is a synonym for *contact. Gain* is a synonym for *profit. Correct* is a synonym for *accurate.* Documentation best practices in my path require me to practice consistency with respect to terms. However, if a word is overused, that can present reading challenges. In such a case, I recommend that you utilize a high-quality synonym.

Tip 352: Writing Business Proposals

Over the span of my twenty-plus year career in Technical Communications, the greatest lesson of all times is this:

Do not reinvent the wheel if the wheel has already been invented. Rather, put the wheel to use.

Business proposals are not something new, so you do not need to pioneer this type of document. My suggestion is that you look to industry-standards and write business proposals from within this established framework. Businesses within a specific industry have a certain range of expectations and in light of this they know how to react and interact favorably. So, when you put together a business proposal that conforms to industry best practices, then your potential client/customer will respond favorably, in most cases.

Sales Managers have asked me on numerous occasions to proof their documents before hitting the road. Like any other document, strong presentation and strong English grammar make for a strong sales pitch and hence business proposal.

Tip 353: Writing Tip for Copywriters

Copywriting is the practice of integrating product knowledge with sales and marketing materials. Copywriters are writing with a sales pitch in mind but must also have sound product knowledge.

My one suggestion/tip for copywriters is this: work closely with product experts to ensure that your copy/text is factually-correct. Then, you endeavor to make sure the document's grammar is correct and appropriate.

Tip 354: Writing Employee Contracts

Infrequently, Technical Writers will find themselves working with in-house attorneys and paralegals. One document that often requires input and review from both legal experts and tech writers is the classical "employee contract". This document details all parameters of an employee's role, working conditions, and relationship with the employer.

Usually attorneys and paralegals write the rough draft for this document, and then ask the in-house tech writer to proof the copy. After necessary revisions, the document can then be used by the employer.

Tip 355: Writing End-User Guides

Technical Writers in the software industry will spend the lion's share of their time writing end-user guides. You see, it is important that the company's proprietary system be used fully and properly. Otherwise, the system will have no client base and the company will close from lack of revenue/patronage.

Technical Writers, therefore, are key business assets in the sense that they transform multimillion dollar software development investments into understandable platforms and apps that patrons can use successfully toward their own business and/or personal ends.

Tip 356: Writing Instructional Guides

Instructional training guides are the main deliverable of Computer Software Trainers. There are two primary ways to present training materials: PPT presentations, and Instructional Design presentations using software like Adobe Captivate and Camtasia Studio.

Tip 357: Writing Job Descriptions

HR departments have the difficult and arduous task of working with diverse managers and departments in an organization in order to write sound *job descriptions.*

A job description is a critical legal document because it states, clearly, the scope and depth of service obliged by the employee and/or contractor. This worker can know, directly, from the job description, exactly what his or her responsibilities are and how to manage the work day in order to accomplish the tasks stated in this official job description. Technical Writers are infrequently asked to help HR staff to write and proofread job descriptions.

Make sure you choose words carefully when composing or editing job descriptions.

Tip 358: Writing LinkedIn Reviews

Technical Writers can help other working professionals greatly on business networking sites like *LinkedIn.* Technical Writers who work side-by-side with talented individuals can take a few minutes out of their schedule to not only connect with but also write a positive review regarding the professional performance of such individuals. The world is a small place, some have told me. This is true. To see a positive review of someone at LinkedIn is a great thing. It means that someone noticed that person's hard work. A LinkedIn review and/or recommendation means that this person deserves to be recognized not just now, but also in the future when this recommendation will play a part in helping the individual to find solid opportunities in his/her field.

Positive energy is like a closed circuit. When the circuit is closed, energy flows well, through and through.

Tip 359: Writing Product Labels

Some Technical Writers will be tasked with the job of writing product labels. Perhaps you work for a Paint Company and your Technical Product Summary (e.g. description of the chemicals/ingredients that comprise the paint) will be placed on the can's external label for the customer to read. Here, there can be zero error. How will you accomplish this task? If I was in this position, I would coordinate the task with my manager. First, I would work with him/her to map out milestones as I gather info and begin compiling and including individual/specific chemicals/ingredients into the label.

Second, I would meet with SMEs and other product experts. Third, I would demand sign-off/approval by not only SMEs but also by key managers and business stakeholders to ensure that the work is not only accurate, technically, but also compliant with industry mandates and standards. In this way, my Technical Writing task is accomplished successfully.

Tip 360: Writing Reports

There are more reports generated each day for businesses around the world than you can possibly imagine. Technical Writers are often tasked with helping Business Intelligence (BI) professionals to quantify and summarize the data presented in the report.

- Reports can help DBAs make decisions. Reports can help programmers make decisions.
- Reports can help management and business owners make important decisions.
- It is one thing to generate a report. It is another thing to be able to understand the data presented and write a cohesive paragraph or summary for that data and explain what it means and its impact on the business.

Technical Writers need to apply their interviewing skills here, in this work scenario. In this way, key meaning can be attributed to the data and then key decisions can be made. Here is a short list of the types of reports (and respective examples) that are generated daily:

- *Feasibility Reports* (Data-As-Evidence Reports)
- *Formal Financial Reports* (Month-End Reports)
- *Investigative Reports* (Research Data Reports)
- *Laboratory Reports* (Experimental Data Reports)
- *Progress Reports* (Activity Improvement Reports)
- *Test Reports* (System Testing Reports)
- *Other Important Business & IT Reports*

Tip 361: Writing Resumés

I have helped numerous friends and colleagues over the years with their resumés. The purpose of this document is *to help you to land an interview*. With that said, here are the key sections of a resumé, in my professional opinion, that you need to include in order to *motivate* the hiring company to *invite you in* for an interview:

- *Header* (your name/title, city/state, email/phone)
- *Objective* (your professional purpose/intention)
- *Qualifications* (your valuable skills/experience)
- *Work Experience* (your leading work experiences)
- *Education* (your formal studies and training)
- *References* (colleagues who recommend you)

The ideal resumé is one page.

However, it is not easy to include all of your top skills, work experiences, and other details in just one page.

So, with caution, go ahead and create a two-page resumé and hope that the reader has the time and patience to turn the page and read the second page.

Tip 362: Writing Technology Articles

There are many freelance writers who enjoy writing about technology outside of a formal office structure.

Technical Writers should have no problem with this career path if deciding to live as an expat or working from a remote location because you are looking after your parents or a close friend, for example. Freelance work is not as consistent as being a full-time contractor or employee. Still, if you can land a gig with well-established magazines/publications like Inc. Magazine, PC World, or Fast Company, for example, you should be able to produce a steady income for yourself.

Why write technology articles? It is simple. Technology is huge. Technology is always growing and expanding.

Technology's growth is a reason behind the writing of this book. In the 80s, computers went from a hidden corporate mainframe center into living rooms and bedrooms around the country and around the world in the form of personal computers.

I believe that Technical Writing, soon, will no longer be performed by seasoned writers like myself. Everyone will need to learn a bit of Tech Writing. Technology writers can harness specific knowledge of an app, platform, software, or device, for example, and write articles for tech publications in need of content. You can essentially turn your hobby into a paid activity. In 2016, I wrote a technology article for a website called *IT Career Finder*, called "Technical Writing 101." Here is the article's web page, should you decide to read it:

https://www.itcareerfinder.com/brain-food/blog/entry/technical-writing-101.html

Tip 363: Writing White Papers

In my professional opinion, there are two "classes" of white papers. First, there are *academic* white papers. Second, there are *business* white papers.

Academic white papers are about specific topics/issues featuring empirical data and research. In some cases, academic white papers lead to a conclusion or perspective. An example of an academic white paper might be "Electronic Payments Exceed Check Payments, Globally, in 2018."

Academic white papers, generally, are not written to market a specific business product or service.

Business white papers are similar to academic white papers in that they, also, address important issues in business, economic, and/or technology. However, business white papers *include* their products and/or services and show how these products/service potentially can help to solve the paper-identified problem/issue.

An example of a business white paper might be "USB Ports Now Accessible To Hackers." The company writing this white paper might have a powerful hardware security software that safeguards USB ports. So, as the issue is explored in the white paper, the company features its security software, for example, that resolves the matter, right then and there.

Business white papers, generally, are written in order to market a specific product or service.

Tip 364: Writing with Praxis

In college, I studied a book called "Cultural Action for Freedom" written by Brazilian Educator, Author, and Social Activist *Paulo Freire*. This book features a term called "praxis" which is the integration of reflection and action. Many times, we discuss things and begin to formulate plans and approaches but leave things unresolved. Praxis is about the conscious and mindful approach of looking deeply into an issue/topic and creating a solution. Then, you take action.

This is an indirect description of Technical Writing.

There could be a business process that needs to be documented. There could be tools at a company that are used in different parts of this overall process. So, the Technical Writer works with management and team leads in different parts of the process to be formalized.

A technical document is then written that maps out and details each stage of the process: who needs to do what and who needs to use what tools and when.

Praxis means completing the task. Praxis means not leaving things unresolved and continuing in conflict or confusion.

Praxis is a solutions-platform not just for educators, but also for Technical Writers, BAs, Trainers, and other IT professionals who see a definitive and detailed document as a way of increasing productivity, reducing risk and confusion, and streamlining resources. The system being developed solves a problem. Its accompanying documentation (e.g. user guide) shows you how to get the most out of the system. This is a win-win approach.

Tip 365: Writing with T.R.U.S.T.

Technical Writing is a practice of T.R.U.S.T.

T = *Truth*. You are working with facts, not fiction.

R = *Respect*. You respect both sources and readers.

U = *Understanding*. You write clear statements.

S = *Sincerity*. You are empathetic toward the reader.

T = *Tolerance*. You are patient as you write.

Note: I came up with this acronym and concept as I was writing other tips for this book.

I like this idea: T.R.U.S.T.

Trust in and of itself is a great virtue and quality in a person. With that said, truth, respect, understanding, sincerity, and tolerance are also important virtues at work in Technical Writers and other Business/Technology professionals.

The more that you cultivate these virtues/qualities, then the more you will (1) write technical documents successfully and (2) gain positive readership.

Thank You for Reading

Thank you for taking the time to read this book amidst your busy schedule. Today, most people are extremely busy. So, sitting down and reading a book is not for the faint of heart. You should pat yourself on the back and say "good job, self." I have done my best to take the greatest of my work experiences and writing insights and *consolidate* them all into 365 Technical Writing Tips.

Should you have any questions about this book, then please email me at thetechwriterguy@gmail.com and I will respond as soon as I can. Thank you for reading!

Keith Johnson, Author, Senior Technical Writer

https://www.techwriterguy.com

Books by Keith Johnson, Live on the Web

- Windows 10 for Seniors and Beginners
- Windows 10 Pocket Guide
- Windows 10 Productivity Guide
- Google Productivity Guide
- Affirmations Productivity Guide
- Affirmations 2020
- Success Principles and Practices
- Buddhist Pocket Guide
- 101 Power Writing Tips
- 100 OM Meditations

All of these books are at ...

https://www.amazon.com/author/techwriterguy

References

Graphics (Images)

Page 22: *Happy Coffee Creative Commons Image*, CC0 License, URL: https://www.pexels.com/photo/happy-coffee-6347/

Page 61: *Google Home Page*, Snagit Screen Shot (2018).

Page 118: *Ben Image*, taken with my iPhone, (2012).

Graphics (Technical Diagrams)

Page 129: *nTier Diagram*, created by me, Snagit (2018).

Page 132: *B2B Diagram*, created by me, Snagit (2018).

Page 134: *Waterfall Diagram*, created by me, Snagit (2018).

Print Books

Alred, Gerald; Brusaw, Charles; Oliu, Walter; *Handbook of Technical Writing, Ninth Edition*, St. Martin Press, 2009, New York, New York, USA.

Carey, Michelle; Hargis, Gretchen; Hernandez, Anne Kilty; Hughes, Polly; Longo, Deirdre; Rouiller, Shannon; Wilde, Elizabeth; *Developing Quality Technical Information: A Handbook for Writers and Editors, Second Edition*, IBM Press, 2010, Indianapolis, Indiana, USA.

Hunter, G. Shawn, *Small Acts of Leadership*, Bibliomotion, Inc., 2016, Brookline, Massachusetts, USA.

Microsoft Corporation Editorial Style Board, *The Microsoft Manual of Style for Technical Publications, Third Edition*, Microsoft Press, 2008, Redmond, Washington, USA.

Web Books

DITA: http://docs.oasis-open.org/dita/v1.2/cd03/spec/DITA1.2-spec.pdf

Microsoft Writing Style Guide: https://docs.microsoft.com/en-us/style-guide/welcome/

Web Pages

https://en.wikipedia.org/wiki/Active_voice

https://en.wikipedia.org/wiki/Adjective

https://en.wikipedia.org/wiki/Adverb

https://en.wikipedia.org/wiki/Adobe_Acrobat

https://en.wikipedia.org/wiki/Adobe_Captivate

https://en.wikipedia.org/wiki/Adobe_Illustrator

https://en.wikipedia.org/wiki/Adobe_InDesign

https://en.wikipedia.org/wiki/Adobe_Photoshop

https://en.wikipedia.org/wiki/Adobe_RoboHelp

http://www.dictionary.com/e/affect-or-effect/

https://en.wikipedia.org/wiki/Agile_software_development

https://en.wikipedia.org/wiki/Altruism

http://grammarist.com/usage/among-between/

https://en.wikipedia.org/wiki/Antivirus_software

https://en.wikipedia.org/wiki/Application_programming_interface

https://en.wiktionary.org/wiki/appear

https://en.wiktionary.org/wiki/display

https://en.wikipedia.org/wiki/Business_rules_engine

https://en.wiktionary.org/wiki/archive

https://en.wiktionary.org/wiki/database

https://www.flhsmv.gov/safety-center/arrivealive/

https://en.wiktionary.org/wiki/artificial_intelligence

https://www.ibm.com/watson

https://ai.google

https://www.microsoft.com/en-us/ai/default.aspx

https://en.wiktionary.org/wiki/aspire

https://en.wiktionary.org/wiki/inspire

https://www.audacityteam.org/

https://en.wiktionary.org/wiki/audience

https://en.wiktionary.org/wiki/authority

https://en.wiktionary.org/wiki/B2B

https://en.wiktionary.org/wiki/B2C

https://en.wiktionary.org/wiki/confidence

https://en.wiktionary.org/wiki/binary

https://en.wiktionary.org/wiki/paragraph

https://en.wiktionary.org/wiki/blogging

https://bookboon.com/

https://www.amazon.com/Business-Speed-Thought-Succeeding-Digital/dp/0446525685

https://en.wiktionary.org/wiki/jargon

https://www.atlassian.com/software/hipchat/downloads

https://www.techsmith.com/tutorial-camtasia.html

https://en.wiktionary.org/wiki/contradiction

https://en.wiktionary.org/wiki/circle

https://en.wiktionary.org/wiki/encircle

https://en.wiktionary.org/wiki/click

https://en.wiktionary.org/wiki/select

https://www.amazon.com/Code-Language-Computer-Hardware-Software/dp/0735611319

https://www.learner.org/workshops/primarysources/revolution/docs/Common_Sense.pdf

https://en.wiktionary.org/wiki/computer_programming

https://en.wiktionary.org/wiki/configuration

https://en.wiktionary.org/wiki/customization

https://en.wiktionary.org/wiki/conjunction

https://en.wiktionary.org/wiki/consistency

https://en.wiktionary.org/wiki/convey

https://en.wiktionary.org/wiki/manage

https://en.wiktionary.org/wiki/corroborate

https://en.wiktionary.org/wiki/collaborate

https://en.wiktionary.org/wiki/courage

https://en.wikipedia.org/wiki/Continual_improvement_process

https://en.wiktionary.org/wiki/craft

https://www.createspace.com/

https://templates.office.com/

http://www.docs.is.ed.ac.uk/skills/documents/3571/3571.pdf

http://www.opentip.org/

http://www.libs.uga.edu/ref/chicagostyle.pdf

https://en.wikipedia.org/wiki/Customer_relationship_management

https://en.wikipedia.org/wiki/Data_type

https://en.wikipedia.org/wiki/Data_management

https://en.wikipedia.org/wiki/Darwin_Information_Typing_Architecture

https://en.wikipedia.org/wiki/Microsoft_SQL_Server

https://en.wikipedia.org/wiki/List_of_DOS_commands

https://en.wikipedia.org/wiki/Document_collaboration

https://en.wiktionary.org/wiki/frustration

https://en.wikiversity.org/wiki/Technical_writing_Document_Design

http://www.ubergizmo.com/how-to/recover-ms-word-document/

https://en.wikipedia.org/wiki/Scope_creep

https://www.pcmag.com/roundup/306323/the-best-cloud-storage-providers-and-file-syncing-services

https://en.wikipedia.org/wiki/System_migration

https://www.investopedia.com/terms/c/core-assets.asp

https://en.wiktionary.org/wiki/permanence

https://en.wiktionary.org/wiki/onboarding

https://en.wiktionary.org/wiki/audit

https://en.wikipedia.org/wiki/AS9100

https://en.wikipedia.org/wiki/Financial_Industry_Regulatory_Authority

https://en.wikipedia.org/wiki/Transparency_(human%E2%80%93computer_interaction)

https://en.wikipedia.org/wiki/Source_code

https://en.wikipedia.org/wiki/Web_portal

https://en.wikipedia.org/wiki/John_Dewey

https://en.wiktionary.org/wiki/efficacy

https://en.wiktionary.org/wiki/efficiency

https://en.wiktionary.org/wiki/enterprise

https://en.wiktionary.org/wiki/error

https://en.wiktionary.org/wiki/email

https://en.wiktionary.org/wiki/employee

https://en.wiktionary.org/wiki/employee

https://en.wiktionary.org/wiki/end_user

https://en.wiktionary.org/wiki/exercise

https://en.wiktionary.org/wiki/experience

https://en.wiktionary.org/wiki/flow_chart#English

https://en.wiktionary.org/wiki/forward

https://en.wiktionary.org/wiki/freemium

https://en.wiktionary.org/wiki/functionality

https://en.wiktionary.org/wiki/glossary

https://www.amazon.com/Google-Productivity-Guide-Charles-Johnson/dp/1502957736

https://www.google.com/drive/

https://www.google.com/docs/about/

https://www.google.com/

https://translate.google.com/

https://en.wiktionary.org/wiki/granularity

https://en.wiktionary.org/wiki/hacking

https://en.wiktionary.org/wiki/hardware

https://en.wiktionary.org/wiki/software

https://en.wiktionary.org/wiki/health

https://en.wiktionary.org/wiki/highlighting

https://en.wiktionary.org/wiki/homophone

https://en.wiktionary.org/wiki/Homonym

https://en.wiktionary.org/wiki/HTML

https://www.programming-book.com/html5-pocket-reference-5th-edition

https://en.wiktionary.org/wiki/index

https://en.wiktionary.org/wiki/interview#English

https://en.wiktionary.org/wiki/laptop

https://en.wiktionary.org/wiki/desktop_computer

https://www.libreoffice.org/

https://www.linkedin.com

https://en.wiktionary.org/wiki/machine_learning

https://en.wiktionary.org/wiki/leadership

https://en.wiktionary.org/wiki/manual

https://en.wiktionary.org/wiki/meeting

https://www.office.com/

https://support.office.com/en-us/excel

https://support.office.com/en-us/outlook

https://support.office.com/en-us/article/microsoft-planner-help-4a9a13c6-3adf-4a60-a6fc-15c0b15e16fc

https://support.office.com/en-us/powerpoint

https://support.office.com/en-us/sharepoint

https://support.office.com/en-us/skype-for-business

https://support.office.com/en-us/visio

https://support.office.com/en-us/word

https://www.amazon.com/Windows-Productivity-Guide-Keith-Johnson/dp/1329889762

https://www.amazon.com/Windows-Seniors-Beginners-Keith-Johnson/dp/153284445X

https://windows-movie-maker.en.softonic.com/

https://docs.microsoft.com/en-us/style-guide/welcome

https://en.wikipedia.org/wiki/Microsoft_Visual_Studio

https://www.lukew.com/ff/entry.asp?1071

https://en.wikipedia.org/wiki/Open-source_software

https://en.wikipedia.org/wiki/Proprietary_software

https://en.wiktionary.org/wiki/zen

https://en.wiktionary.org/wiki/neither

https://notepad-plus-plus.org/

https://en.wikipedia.org/wiki/Note-taking

https://en.wiktionary.org/wiki/noun

https://en.wiktionary.org/wiki/organization_chart#English

https://en.wiktionary.org/wiki/outline

https://en.wiktionary.org/wiki/pagination

https://en.wiktionary.org/wiki/paragraph

https://en.wiktionary.org/wiki/parallelism

https://en.wiktionary.org/wiki/paraphrase

https://en.wiktionary.org/wiki/passive_voice

https://en.wiktionary.org/wiki/part_of_speech#English

https://en.wiktionary.org/wiki/PDF

https://en.wiktionary.org/wiki/percent

https://en.wiktionary.org/wiki/period

https://en.wiktionary.org/wiki/phenomenon

https://en.wiktionary.org/wiki/phrase

https://en.wiktionary.org/wiki/plagiarism

https://en.wiktionary.org/wiki/plan

https://en.wiktionary.org/wiki/platform

https://en.wiktionary.org/wiki/priority

https://www.wikihow.com/Make-a-Process-Document

https://www.apm.org.uk/resources/what-is-project-management/

https://en.wikipedia.org/wiki/Punctuation

https://en.wikipedia.org/wiki/Requirements_analysis

https://en.wikipedia.org/wiki/Trust,_but_verify

https://www.techsmith.com/screen-capture.html

https://en.wikipedia.org/wiki/Search_engine_optimization

https://en.wikipedia.org/wiki/Software_development_process

https://en.wikipedia.org/wiki/Standard_operating_procedure

https://www.stc.org/

https://www.amazon.com/Success-Principles-Practices-Keith-Johnson/dp/1537377167

https://www.grammarly.com/blog/verb-tenses/

https://en.wikipedia.org/wiki/Test-driven_development

https://en.wikipedia.org/wiki/Glocalization

https://en.wiktionary.org/wiki/teamwork

https://www.udemy.com/

https://www.udacity.com/

https://en.wikipedia.org/wiki/User_interface

https://en.wikipedia.org/wiki/User_experience

https://en.wikipedia.org/wiki/Gerund

https://en.wikipedia.org/wiki/Iteration

https://en.wikipedia.org/wiki/Multitier_architecture

https://en.wikipedia.org/wiki/Scalability

https://en.wikipedia.org/wiki/Use_case

https://en.wiktionary.org/wiki/unity

https://en.wiktionary.org/wiki/user_guide

https://en.wiktionary.org/wiki/transitive_verb

https://en.wiktionary.org/wiki/intransitive_verb

https://en.wikipedia.org/wiki/Waterfall_model

https://www.google.com/chrome/

https://www.microsoft.com/en-us/windows/microsoft-edge

https://en.wikipedia.org/wiki/Whiteboard

https://en.wikipedia.org/wiki/Website_wireframe

https://en.wikipedia.org/wiki/Acronym

https://en.wikipedia.org/wiki/Opposite_(semantics)

https://en.wikipedia.org/wiki/Stakeholder_(corporate)

https://en.wikipedia.org/wiki/Subject-matter_expert

https://en.wikipedia.org/wiki/Synonym

https://www.itcareerfinder.com/brain-food/blog/entry/technical-writing-101.html

https://en.wikipedia.org/wiki/Praxis_(process)

https://en.wikipedia.org/wiki/White_paper

www.ingramcontent.com/pod-product-compliance
Lightning Source LLC
Chambersburg PA
CBHW071305220526
45468CB00001B/280